PACKAGING SOURCE BOOK

PACKAGING SOURCE BOOK

ROBERT OPIE

Macdonald Orbis

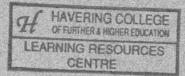

A Macdonald Orbis **BOOK**

Copyright © 1989 Quarto Publishing plc

First published in Great Britain in 1989
by Macdonald & Co (Publishers) Ltd
London and Sydney

A member of Maxwell Pergamon Publishing Corporation plc

British Library Cataloguing in Publication Data
Opie, Robert 1947-
   1. Packaging source book.
   I. Title
   688.8

ISBN 0-356-17665-7

This book was designed and produced by
Quarto Publishing plc
The Old Brewery, 6 Blundell Street
London N7 9BH

Senior Editor
David Game

Editor
Susan Berry

Designer
Hazel Edington

Photographer
Rose Jones

Art Director
Moira Clinch
Editorial Director
Carolyn King

Typeset by Ampersand Typesetting (Bournemouth) Ltd
Manufactured in Hong Kong by Regent Publishing Services Ltd
Printed in Hong Kong by Leefung-Asco Printers Ltd

Macdonald & Co (Publishers) Ltd
Headway House
66-73 Shoe Lane
London EC4P 4AB

*Sapporo Draft Beer (1988). The can has
a ring pull which removes the entire
lid making it into an effective
drinking vessel.*

CONTENTS

INTRODUCTION

1

Matches needed a box not only to hold a quantity together safely, but also to provide a convenient surface on which they could be stuck. Friction matches were first made in 1827 and within thirty years many European countries were producing them. The London provision merchants Bryant & May had imported matches from Sweden since 1852 and started to manufacture their own in 1861. By the 1870s the match box label had become colourful and decorative, such as Bryant & May's Tiger Match (1). On many brands the picture alone sufficed, and the brand name itself did not appear.

Emerson observed that, "The craft of the merchant is the bringing a thing from where it abounds to where it is costly". When Emerson made this remark, more than a hundred years ago, he was thinking principally of the problem of transportation. The particular product he had in mind (the value of which would increase a hundredfold if it could be carried from the farmer's tree) was the American peach.

Although the packaging of fancy goods was already commonplace by the time Emerson's book, *The Conduct of Life*, was published in 1860 – and the packaging itself was often fanciful, with coloured embossed papers, romantic lithograph pictures, and finely engraved labels – the retail packaging of perishable goods was still in its infancy. Only in the past hundred years has it become possible, for instance, to can peaches in the vicinity of the orchards where they are grown on the day they are picked. As a result, efficiency of transportation has ceased to be the producer's main concern, and a new range of considerations now dominates the marketing of consumer goods. The basic functions of the sealed package – to protect the product, to enhance its appearance and to facilitate its distribution – were soon to be matched by others, more subtle perhaps, but no less far-reaching in their consequences. In the middle years of the nineteenth century the provision merchant had often been charged with the adulteration of foodstuffs, with the giving of short weight and, if he were the sole supplier, with profiteering. When John Horniman began packing his blended teas in sealed fixed-price packets that bore his name and address, with the guarantee "Pure Mixed Tea – Full Weight Without Package", he found grocers unwilling to retail his product. He was forced instead to sell through chemists' shops because his pre-packaged teas undercut the prices the grocers were charging for blending teas themselves from bulk purchases.

The benefit of the manufacturer's package is that it passes on to the consumer the advantages of mass production, mechanical handling and, in some ways, of

2

In 1823 Lea & Perrins set up business in Worcester, England, as chemists and druggists. By chance they had acquired a recipe for a sauce that had originated in Bengal, India. In 1837 they launched Lea & Perrins' Worcestershire Sauce. The orange label with its border that continuously repeated the words Lea & Perrins was devised in 1840. Modified from the original, the label here (2) dates from the 1930s. The glass stopper was replaced in 1957 by a plastic pouring fitment.

bulk buying, no matter how small each individual purchase. The manufacturer, rather than the retailer, is therefore responsible for the quality, the quantity and the definition of the product. And most significantly, it puts the consumer, no matter now humble, in direct contact with the manufacturer.

Once the pre-packaging of goods became general, the development of the self-service store was obvious. The well-designed package is, or should be, its own salesman. To have preserved the old style of retailing would have been to have ignored the services of the thousands of silent salesmen newly marshalled in each store. Yet the change is so radical, and has come about so recently, that it

4

3

Most brands retained their familiar profile for as long as possible, such as with Lea & Perrin's Worcestershire Sauce (2). The pack design became part of the product.

However, in the 1930s some brands, such as Rinso (3), found it necessary to draw their customers' attention to certain advantages of their product, and announced them boldly on the front of the pack.

would be a wonder if all the implications of the package as a marketing medium were yet fully appreciated. The package is now the object that the shopper looks for and hopes to recognize in a supermarket. It is also the article that he or she picks up and has to carry home. And it is the package that the purchaser subsequently lives with for a while, looks at, consults in time of need, helps himself from, discards, and then recalls with pleasure or distaste. If McLuhan's dictum that in communications "the medium is the message" is accepted, then in merchandizing it might now be said "the package is the product".

However central the style and design of a package is to a sales campaign, it is still simply a container. To many people its colour, its cleverness as a dispenser, its provision of free reading matter, or its ability to turn into a yo-yo after it has been emptied, are superfluous, and maybe a source of irritation. Not everyone is convinced that contemporary marketing methods are the best, or for the best. The necessity of packaging (and for that matter, promotion as well) is questioned. The argument appears irrefutable that commodities that are essential to life will continue to be purchased no matter how economically packaged, and whether or not promoted. But it is the competitive market place that ensures the need for products competitively priced, carefully packaged and creatively promoted.

Certain colour combinations work well together: yellow and red, for instance, were often used by British tobacco manufacturers for their Gold Flake variety. Indeed the colour combination became associated with that type of tobacco. In France the successful yellow and red combination was adopted by manufacturers of cooking stock: Kub, Maggi and Poules au Pot (4).

INTRODUCTION

Some of the earliest product designs appeared on labels for beers and spirits. The triangular trade mark label for Bass (1) was used from 1855 and became familiar around the world, as did the Guinness harp design from 1862. The Martell (2) label was used from 1848 and Hennessy's from 1865. All these labels survived unscathed until the supermarket era, and Martell's lasted out until 1968.

Once the contents were safely sealed in their container, it was important that their nature could be readily identified. (For the ancient Romans, the shape of an earthenware pot indicated whether wine or water was within.) In the case of wine, an impressed mark might also indicate from which vineyard it had come. By the eighteenth century printed paper labels started to appear on glass phials for drugs and on wine bottles, as yet with little decorative element. But in the same century, more elaborate typographic and pictorial designs began to appear on wrappers for tobacco and for pins, as well as on more substantial boxes such as those for gloves, where the label was often more of an extension of the manufacturer's trade card. Such designs were the province of the engraver in copper, steel and wood, whose craft enriched the visual appeal of the tradesman's wares as well as his bill heads and business cards.

As the techniques of commercial printing improved, so too did the quality of the commercial engraver's art. By the 1850s colour printing was able to enhance many an image, and during the next thirty years designs for a wide variety of commodities were created, particularly for spirits, tobacco, sauces, toiletries and medicines. Many world-famous images first appeared during this time, notably those for Bass (1855), Guinness (1862), Martell (1848) and Hennessy (around 1860).

1

2

3

Victorian design was often ebullient, frivolous, detailed and colourful. Such delightful "jewels" on the grocer's shelf must have added greatly to the appeal of more mundane products like washing soap or health salts. In the medicinal area, the packaging was less colourful, traditionally black printing on a white label. As colour packaging developed, many medicinal brands were provided with colourful wrappers or boxes to make them more attractive, and yet they still retained their familiar, respected but dull image inside.

In the case of Bird's Custard (an egg-free product invented by Alfred Bird in 1837 to enable his wife – who was allergic to eggs – to eat custard), the packets were sold through chemists' shops until the product became more widely accepted. So that it might compete better with the more colourful array of products on the grocers' shelves the design was transformed around 1900 into the style which it retains to this day.

As manufacturers launched more and more products on to the market, the need to establish each brand's presence and image in the mind of the public became increasingly important. Much of this was done through advertisements in newspapers and magazines, and on posters. Over a period of time the brands became so familiar that purchasers regarded them as friends. Indeed the product's image, created by its promotional slogans and epitomized in the pack's design, became the trusted visual symbol for each product.

It was not long, however, before manufacturers found they had a dilemma. New brands arriving at the beginning of the twentieth century were being decked out in the modern styles of the time, for example in fashionable Art Nouveau. By comparison the earlier designs started to look out-dated and it was felt that this would adversely affect their reputation. The dilemma confronting manufacturers was: should the established brands with their familiar, recognizable design, so carefully promoted and nurtured over many years, be changed? In many instances, designs began to alter subtly, but not by enough to cause the buying public to wonder whether the contents themselves had changed.

By the 1930s the intricate designs of former years were fast disappearing and by the 1950s, with the coming of the self-service store, the design emphasis moved totally on to a clean-cut image with shelf-appeal. The product could no longer rely upon the grocer's recommendation; it had to proclaim itself loudly among the ranks of its competitors.

In recent years, Britain has led a renaissance in package design; more radical changes are now being made than at any previous time. A particularly noticeable trend in package design is the "own label" products of the supermarket chains. They have less perhaps to lose than the established manufacturers, as well as having a following of regular customers who prefer their slightly cheaper products.

Apart from the overall design, the two elements that contribute most to the product's visual image are a prominent colour (or colour combination) and, where

INTRODUCTION

1

USA), had just this kind of impact. A similar style had been used by Oxydol in the 1930s and a sun-burst motif had been used to good effect by Reckitt & Colman when they launched three famous products around the turn of the century.

Packaging materials

It was not until the beginning of the nineteenth century that the characteristics of commercial packaging – branded, colourful and individually packed – had evolved. At that time glass bottles were available for such luxuries as perfumes, sauces and wines, some of which had hand-printed paper labels. Pots for ointments were in common use while stoneware bottles contained all manner of cheap liquids, even shoe blacking. Dry products like tea and tobacco were wrapped in paper by the grocer or tobacconist. Box-making from wood or tin was in its infancy, the latter being mainly used for snuff.

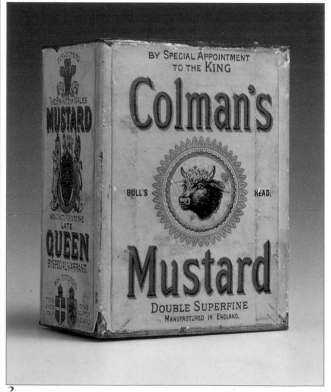

2

Of all the packaging materials available before the arrival of plastic, glass offered the most effective way of creating a distinctive and individual shape. Three brands that have made their mark with shaped bottles have been, Bovril (1), the meat extract launched in Britain in 1886 by a Scotsman (having first been successfully tested in Canada); Perrier, the French table water, started by an Englishman in 1903; and Orangina in 1937.

possible, a distinctive shape. With many products the variety of shape is restricted by the materials used, but with glass (and now plastic) some brands command instant recognition through shape alone. Coca-Cola and Orangina are good examples of brands of soft drink that have relied on a distinctive bottle shape to give them such a clear identity that little additional labelling is required.

Pack colour has a vital role in a products' immediate recognition. Before the actual design registers, the customer becomes aware of the overriding colour. Prominent examples are the yellow of Colman's Mustard and the red of Royal Baking Powder. In the case of Cadbury's products, the purple colour has become synonymous with a whole range. When brands have a powerful colour link, it becomes easier for design changes to be implemented, so when images were simplified, the background colour was often accentuated as well.

A prominent pattern which totally dominates the pack can also produce an instant impact. When Tide was launched in 1950 the carton, with its concentric rings in electric orange and yellow (the style is still in use in the

The firm of J & J Colman of Norwich, England, was founded in 1823 to make mustard, and the bull's head trade mark has been used since 1855. The use of mustard yellow for the label was an obvious but successful choice. By the 1870s the classic style of the

Colman's Mustard design (2) had probably already developed. Today's pack has a similar feel to the tin shown here dating from around 1905, although the design was changed for a time in the 1960s.

During the nineteenth century the packaging of products developed quickly. In the environment of the industrial revolution, it was the combined efforts of the inventor, the mechanic and the entrepreneur that became the driving force of the packaging age.

The development of cheaper printing during the first thirty years of the century greatly stimulated the packaging industry, making it cheaper and easier to wrap products; but what really transformed it during the second half of the century was the coming of cheap colour printing, which transformed the humble tin box, the labelled bottle and the simple carton into things of beauty.

Some products were not affected by the new developments in packaging. The stoneware bottle used for dispensing ginger beer, for instance, remained relatively unchanged until its demise in the 1930s. Surprisingly, the earthenware pot later developed into an item of luxury status as transfer-printed decorations adorned the pot lid, first in black on a white glazed surface (to promote tooth powder or bear's grease, used to increase the growth of hair) and then, in the 1850s, in multi-coloured form. The best-known company producing these lids was R. Pratt & Co of Staffordshire. Although the pots contained meat paste or potted shrimps, they were often bought as much for their pictorial lid (which made no mention of product or manufacturer) as for their contents. By the end of the century the popularity of the pictorial pot lid had waned, perhaps because by that time so many other products were as colourful.

One major advantage of the traditional glass container was that the buyer could see the contents and, if they were colourful and exotic, found it much more tempting. Glass could also be moulded into all manner of elaborate shapes and decorations. In America, the moulding of glass whisky flasks became an art form from the 1820s onwards,

Around 1900 the sun-burst motif was popular. It was used for a number of brands by the Hull-based firm of Reckitt & Sons, who made household products. Robin Starch (3) was launched in 1899 in five pack sizes with the enticement that "the iron does not stick to the linen but glides over the surface". In 1905 another sun-burst arrived with Brasso, Britain's first major liquid metal polish. The tin retains the same design today. The liquid grate polish, Zebo, was launched in 1921.

INTRODUCTION

*The firm of L Rose & Co was established at Leith, Scotland, in 1865. Various lime juice drinks were sold, including a lime juice champagne. Rose's Cordial (**1**) became the most popular, prepared from West Indian limes and "entirely free of alcohol". During the 1870s the glass bottle had developed a distinctive embossed pattern of lime branches, but it did not take long for competitors to imitate the style.*

1

with rival glass manufacturers competing to make the best impressions: for example, the portrait of George Washington. In Britain, the most enduring, and perhaps endearing, moulded bottle was that for Rose's Lime Juice, probably devised in the 1860s, which continued (although with less pronounced embossing) to be manufactured until 1987 when it was finally replaced with plastic.

A special type of glass bottle was developed for aerated waters at the beginning of the nineteenth century. Artificially produced mineral waters had been made since the 1770s, but the pressure of the carbonated drink tended to escape unless the bottle was extremely well corked. The answer lay with a bottle (developed in 1814) that had a pointed end. Known as a "Hamilton" after its inventor, this bottle had to be laid on its side to keep the cork moist and tight fitting. No improvement of any significance was made in this area until, in 1872, Hiram Codd of London perfected a new bottle stopper: a glass marble held tight inside the opening by the pressure of the effervescent drink. In the same year, the internal screw stopper was also invented, a simple device that could easily be used to reseal the bottle and which remained popular throughout the world for over a hundred years. Another equally popular form was the swing stopper (as used on Corona bottles and Grolsch beer bottles) invented in 1875; again, it was easy to reseal firmly.

In 1800 France and England were still at war, and it was during this time that Napoleon realized that his army needed to have better food on their campaigns: "an army marches on its stomach". He offered a prize to anyone who could come up with a solution – portable nourishing food that would keep indefinitely. It was Nicolas Appert who produced, in 1909, the means of cooking meat in a glass jar sealed with a bung stopper. The vacuum created preserved the contents until the jar was opened, a method that still holds good today. Soon after, this principle became known in Britain where tinplate containers were first used – although only in limited quantities – by the British Navy and by polar explorers. The idea of canned food was slow to be accepted by the public, and even in the 1920s it was still frowned upon by most.

In America, metal canning came into use from 1837 onwards when glass jars became too expensive for many items. During the American Civil War cans were used extensively and this hastened their acceptance in the home. By the 1880s canning factories had spread across America, and were exporting canned fish, fruit, vegetables, condensed milk and, of course, vast quantities of corned beef in the characteristically tapered-shape can with its key opener.

Tin boxes were being used in the first half of the nineteenth century for a number of grocery items, in particular for biscuits, for which the market was growing in Europe. Increased mechanization of biscuit production in the 1850s produced a large surplus to local needs, and this meant that the fragile biscuit had to cope with the perils of wider distribution at home and abroad. The tin box provided the best means of preserving the biscuits; it was strong enough to prevent breakage and, when sealed, kept them fresh. (Huntley & Palmer had used tins since the 1830s.)

Fancy biscuits were something of a luxury in Victorian Britain, and, to increase their sale at Christmas time, manufacturers packed their special biscuits into increasingly decorative tins. With the ability to print colour directly on to tin – Huntley & Palmer first did this with a transfer print in 1868 – the era of the Christmas biscuit tin had begun. Printing by offset lithography soon followed, with even better results. It also became possible to fashion the tin into an amazing array of fancy shapes which, by the beginning of the twentieth century, could resemble, for example, a bird's nest, a snakeskin wallet, or even a row of books.

The success of the biscuit tin must have encouraged other manufacturers to make use of the medium. Mustard, cocoa and tobacco firms were all early users of tin boxes. In America where bakers supplied only local needs, the use of tin boxes was much more limited until the 1880s, biscuits, for example more often being packed in wooden boxes and barrels. However, the American public was being persuaded to buy cartons of cookies (rather than buying the biscuits loose) long before the British public were forced to do so by the arrival of the self-service store. The versatile carton and paper wrapper catered for the cheaper end of packaging. Indeed traditionally it was the shopkeeper who would have a good supply of wrappers to hand, ready to wrap whatever commodity his customers chose to purchase. During the nineteenth century most purchases – tea, sugar, cereals, dried fruit and so on – were delivered by the manufacturer to the retailer in bulk and had to be weighed out and then wrapped for each customer – the grocer was most dextrous at conjuring up his own bags or twists.

In the 1850s some paper manufacturers produced ready-made paper bags which gradually began to play a major, albeit humble, role in the story of retail packaging. The first paper bag-making machine was developed by Francis Wolle of Pennsylvania, USA, in 1852. It was not until 1873, when the idea was picked up by a visiting Englishman, Elisha Robinson, that the technology arrived in Europe. By 1902 the firm of E.S. & A. Robinson of Bristol had seventeen bag-making machines, although 400 people were still being employed by the firm for bag-making by hand.

The skill of making cardboard boxes was being practised in the early years of the nineteenth century in England, France and America. There was a demand for everything from small pill boxes to large hat boxes. In England by the end of the 1850s, the firm of Robinsons of Chesterfield advertised their ability to produce over 300 different sizes of box. Some years later, there was to be a great demand for a new Christmas treat, the fancy chocolate box. In 1868 the confectionery firms of Fry's and Cadbury's both brought out a small range of pictorial boxes filled with chocolates, for the delectation of the British public.

The demand for cardboard boxes rose steadily, but if the box was going to succeed in the grocery store it had to be cheap and mass-produced. The answer lay in the folding carton, by which means a complete box could be made by cutting and creasing a single sheet of card. It had one other major advantage – when stored flat (before being filled) the carton took up little space. The first folding cartons (for carpet tacks) are thought to have been made in America around 1850 but the carton was destined to play a major role in the packaging story, especially when the addiction to cigarettes became widespread.

George Palmer joined Thomas Huntley's biscuit business at Reading, England, in 1841. Large 10lb tins were used to distribute the biscuits around the country and to export them to every corner of the world. The Huntley & Palmer label (2) most frequently associated with the tins was the one devised in 1851 which employed a buckle and garter. It was still in use a hundred years later.

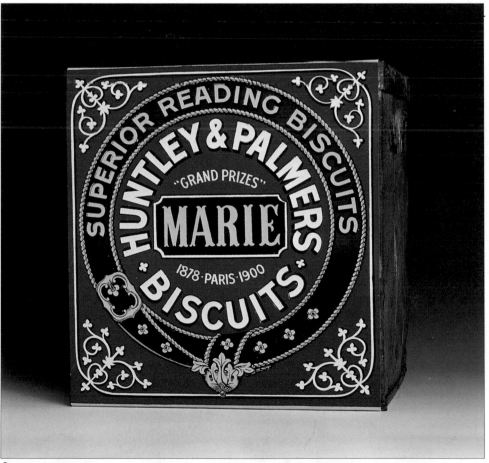

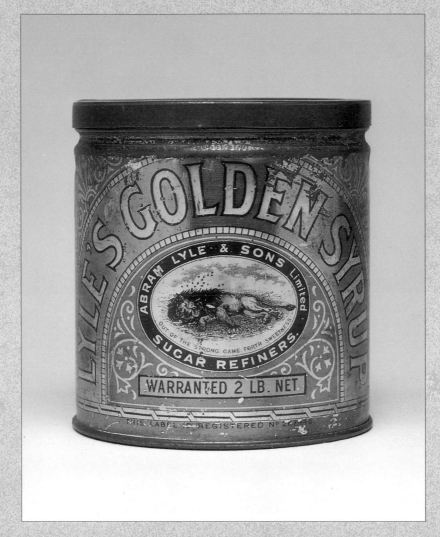

In 1883 Abram Lyle sent his five sons from Greenock in Scotland to London to build a sugar refinery. After a difficult start, they had established a refinery that also produced a treacly syrup, which was refined further into a saleable product. Lyle's Golden Syrup went on sale in 1885 with the lion and bees trade mark, and bearing the legend "Out of the strong came forth sweetness".

CHAPTER ONE
CLASSIC PACKS

1880-1899

INTRODUCTION

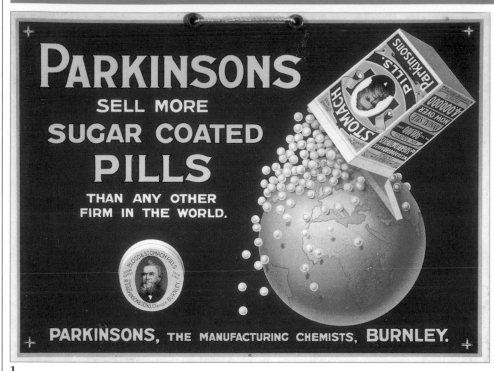

1

The packaging for Parkinson's medicinal products looks almost gaudy compared with the subdued colours of most other late nineteenth-century lozenge, pill and ointment packs. The firm of R. Parkinson & Sons started in 1848 at Burnley, England. The founder's image appeared on many of his products, and by the 1880s the advertisements (1) proclaimed that Parkinsons sold more sugar-coated pills than any other firm in the world.

No longer promoting themselves as the manufacturer, they could now focus on a particular brand and advertise it, even seemingly as a "competitor" to the other brands that they produced. However, this was not the most significant change that the pre-packed brand brought about. Now that the manufacturer rather than the retailer apportioned the product, recommended its retail price, and wrapped it for the customer to carry away, the manufacturer could make himself known to the individual consumer. The manufacturer labelled the package, placed his name on it, chose the brand name for his product, and could add, if he wished, his own assurance of its quality as well as its other virtues.

2

During this period most of the products that the grocer sold were still being delivered to him in bulk; his tea, flour, sugar, rice, oatmeal, dried fruits and many other dry goods arrived in wooden chests or sacks. He then had to weigh out and wrap up individually the quantity required by each customer. This was both a skilled and a time-consuming job; and it was time-consuming for the shopkeeper, and his customer. But these were more leisurely times (for some at least), and customers were often provided with chairs on which to sit whilst their orders were made up for them.

Nevertheless, branded goods were starting to make their presence felt. Whereas numerous firms supplied products under their own name, such as Johnson's Baby Powder, there were now more products appearing more frequently which had an entirely new name, distinct from that of the manufacturer. Thus it became possible for a range of similar products to be manufactured by the same firm. Tobacco companies were amongst the first to adopt this practice; for example, in the 1880s Wills listed many exotic or romantic names for their tobaccos, such as Honey Flower, Autumn Gold, Evening Star, and the unlikely-sounding Bishop Blaze (introduced in 1847 and named after an unfortunate churchman who was martyred in AD 316). Such brand names gave added appeal to the product. In 1879 Ivory was the name given by Procter & Gamble in America to a hard white soap they produced, and on the other side of the Atlantic, William Lever named his best yellow soap Sunlight.

Since 1849 R. Paterson & Sons of Glasgow had preserved pickles and chutney. In 1885 they produced a liquid coffee essence for use by the Gordon Highlanders who were serving in India at the time. It was called Camp Coffee (2); the label depicted here was similar to the original. In 1957 the servant's tray was removed – causing tradionalists in India to complain.

18

In order to publicize these new brands, massive advertising campaigns on a national basis became commonplace during the last part of the nineteenth century. Posters on the hoardings and advertisements in magazines announced their arrival, extolled their virtues and continually reminded everyone of their existence. Thus the product was, in a sense, already "sold" before the customer entered the shop and recognized the article by the name or design on the package.

These new pre-packed goods, however, did impose a restriction on the customer. Previously, it had been possible to see the product and sample it, but now it was sealed from close inspection. (Manufacturers did often

4

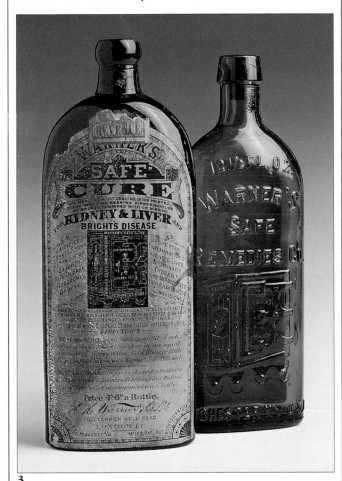

3

Hulbert Warner had built up a business in Rochester, New York State, selling safes, but he then decided to diversify. In 1879 he produced his first medicine, Warner's Safe Cure for Kidney and Liver (3). A safe was the obvious

choice of trade mark. A stream of other "cures" followed, but in 1906 the Pure Food and Drugs Law banned the words "cure" and "safe" as descriptions for medicines.

supply miniature samples of their wares, allowing the public to try new brands.) But now that the product was wrapped, it was this very wrapper that had to persuade the customer to make the purchase, assuring the purchaser of the quality of its content.

The immediacy of the pack design was important; a colourful, bright and exciting image not only attracted the customer but gave the product a feel of hygiene and freshness. In addition manufacturers developed a whole range of devices to give their brands much needed credibility. (For many years the public had been subjected to unscrupulous retailers and producers who had allowed adulterated goods to be sold.) For some fifty years international exhibitions had been held around the world where the latest products were displayed, and medals and diplomas were awarded to those excelling in each category. It was with great pride that manufacturers printed on their packs the occasions on which these medals had been won, and often the medals themselves were incorporated into the pack design.

In Britain there were also city council analysts who would inspect the product and pronounce on its worth. A guarantee could then be printed on the pack. A more direct approach was to offer a reward to anyone who could find fault with the product. This idea was used extensively by Lever Brothers on their packs of Sunlight soap. The then enormous sum of £1,000 was offered to anyone who could prove that the soap "contains any form of adulteration whatsoever or contains any injurious

Those companies that were awarded medals for their wares were naturally very proud of them. In the case of Hignett's of Liverpool, England, the prestigious tin that contained their Cavalier Brand Bright Flake (4) boldly displayed two gold medals. The first was awarded in 1884 at the International Health Exhibition held in London, and the second at the International Exhibition of Navigation, Travelling, Commerce & Manufactures held at Liverpool in 1886.

INTRODUCTION

chemical"; a further guarantee authorized dealers to return the purchase money to "anyone finding cause for complaint".

Of even greater importance to the manufacturer though was royal patronage. This enabled both British and many European companies to emblazon their packs with a royal coat of arms, and royal approval could be no better testimony to the contents. However, in America the face of the product's creator was often incorporated into the pack design, especially for medicinal products. The face that probably became the best known in the twentieth century was that of K.C. Gillette whose portrait appeared on the countless millions of razor blade wrappers sold around the world. The picture of a trustworthy gentleman, a face with which to relate, seems to have given confidence to the purchaser. In Britain the firm of John Oakey & Sons had secured as their trade mark the notable figure of the

2

Duke of Wellington, whose image gave stature to their range of products.

The introduction of the Trade Marks Acts in Britain in the 1870s gave companies legal entitlement to the registering of a symbol that no other firm could use. Before then there was the continued risk that other manufacturers would create cheap imitations of successful brands. It also gave the customer the assurance that the trade mark was genuine and could be trusted. Another way of preventing fraud was for the manufacturer to sign each carton or label – a task made less tedious when it was printed as part of the pack: "None genuine without this signature".

In this way the company's reputation was built up, and reliable firms who consistently made products of good quality and were able to sell them successfully were rewarded with larger orders. As the firms grew, so did the factories, which became the wonders of the industrial age. Their owners were so proud of them that they often featured in their advertisements and sometimes on the container as well.

In the factory, advances in mechanization were making it more practical to mass-produce the containers, and this in turn made it easier for the manufacturer to supply the retail outlets with individual packs. Cadbury's, for instance, had installed a machine that would measure out 12,000 packets of cocoa essence per day, with the further advantage that the contents were not touched by human hand during the whole process of production.

For cigarettes, the machine that revolutionized the market was patented by James Bonsack of Virginia, USA, in 1881; it mass-produced cigarettes that in the past had been rolled by hand. Although his was not the first attempt to mechanize the process, it was more successful than the others; but it was not till 1888 that Wills (who bought the exclusive rights) improved the machine sufficiently to put

1

*The American can label for HK & FB Thurber's roast turkey (**1**) dates from around 1890.*

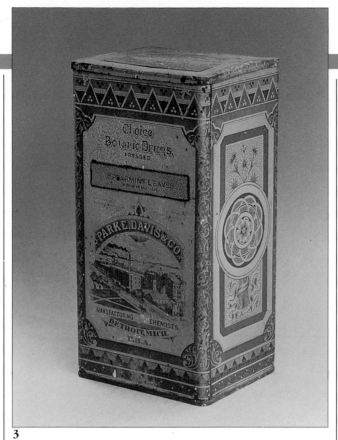

3

Hiram Codd had patented his marble-stoppered bottle and the internal screw top had been devised, others had attempted to find a better idea but few of the hundreds of variations were an improvement. In the USA an exception was the resealable Hutchinson bottle of 1879 in which a metal wire was connected to a porcelain stopper, sealing the bottle from inside. It was only when William Painter of Baltimore patented his Crown cap in 1892 that a successful alternative closure became available.

Its advantage lay in its simple application, a metal disc crimped on to the bottle top. It was a cheap operation, with the added virtue of being effective and tamper proof but its disadvantages were that the bottles could not easily be resealed and that a bottle opener was needed. However, by the 1960s it had replaced the internal screw stopper on beer and soft drinks, although by then aluminium screw tops were winning a share of the market.

it into widespread commercial use with the introduction of Woodbine cigarettes at five for one penny.

Around this time advances were being made in the development of airtight tins, used by the British biscuit and tobacco firms for keeping their products fresh, a necessity when exporting. The most ingenious advance was that made by G.H. Williams. His tin was cylindrical and was opened by revolving the lid, which had an in-built cutter designed to pierce the airtight inner foil. The firm of Wills had exclusive rights to this invention, which they started to use from 1887.

Another metal package with new potential was the collapsible tube. It had been used since the 1840s for artists' colours, first of all in America; but it was not till 1892 that the idea of putting toothpaste into tubes was made practical. Colgate experimented with tubes soon after this and found them acceptable to the public.

In the field of paper packaging a marked improvement was made by Robert Gair of New York in 1879, when he produced a means of speeding up the process of carton folding. British firms were slow to take on his machines, and it was not until the late 1890s that they were being used extensively, principally for the growing cigarette market. The Germans were developing similar machines, and by 1897 some 800 patents relating to folding boxes had been registered worldwide.

Much attention was also being given during the 1880s and 1890s to the development of glass bottles, and particularly to their closures. Ever since 1872 when

4

By the end of the nineteenth century the factory was the wonder of the age. Many manufacturers promoted themselves by displaying pictures of their imposing factories in their advertisements or on their packs (3), as did Parke, Davis & Co, manufacturing chemists from Detroit, Michigan.

Aass Bokk lager beer still uses the old style of label on this modern bottle (4). The label includes the medals won since the 1880s and depicts the brewery, which is Norway's oldest, having been founded in 1834.

SMOKING

Tobacco mixtures could be made from a varied selection. Richmond Club Mixture (**1**) by Cameron & Cameron was made from tobacco grown in Virginia and North Carolina and then blended with Havana, Louisiana, Perique and Turkish tobaccos, to give a "truly delicious, sweet and pleasant smoke".

Tobacco is best preserved when kept in airtight containers, and so tins were used extensively by tobacco manufacturers. The American designs on this page date from around 1890, and the Edgeworth (**2**) design has remained virtually unchanged.

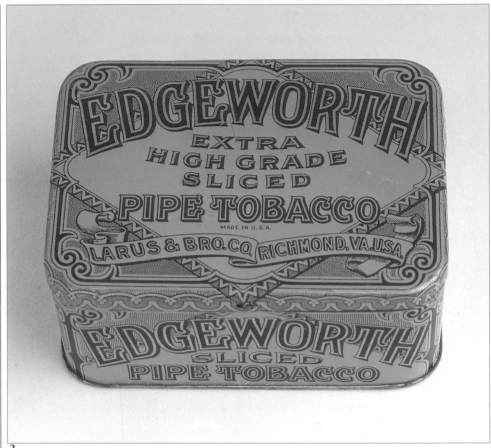

2

1

3

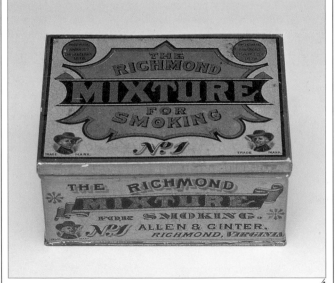

4

These American Cupid Bouquet Little Cigars (**3**) have a local tax stamp to seal the tin, such seals followed the implementation of an act of 1897.

Some manufacturers did not care much about the quality of tobacco but for Allen & Ginter's Richmond Mixture (**4**) there was an assurance for their customers that the tobacco was "entirely free from stems, artificial flavouring or adulteration; it does not swell, burn hot, or pop, nor does it cake or go hard in the pipe, however tightly filled, and burns to the last grain".

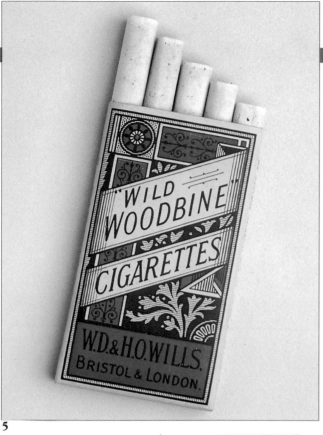

5

Wills' "Wild Woodbine" cigarettes (**5**) were launched in 1888 at the low cost of five for a penny, a price that remained constant till 1915. Cheap cigarettes had been made possible in the UK by the introduction of the Bonsack machine, which was already proving itself in the USA. By 1893 over 150 million Woodbine cigarettes were being sold annually, and they remained as Britain's leading cigarette until 1960. In that year the classic design was changed – in an attempt to arrest a marked decline in sales.

Although there were over 500 registered tobacco firms in Britain in 1880, most were small and only around 100 produced highly decorative tins, the picture usually linked to the brand name, as it was with Kilty Brand (**7**) and After Lunch (**8**).

7

8

Player's Navy Cut and Mixture tobaccos (**6**) were the mainstay of the Nottingham firm. Their sailor's head trade mark of "Hero" had been registered in 1883; the lifebelt was added five years later. Player's Navy Cut cigarettes, which cost 3d for ten, were launched in 1900.

6

The tobacco manufacturers Salmon & Gluckstein also had a chain of retail shops in the London area. By the time they were sold to the newly formed Imperial Tobacco Co in 1902 there were 140 outlets. Their Snake Charmer cigarettes (**9**) were gold tipped.

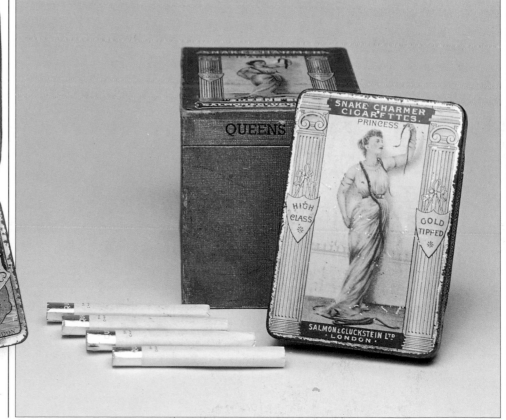

9

LIGHTING UP

Thomas Gee of Gloucester first portrayed a battleship, H.M.S. Devastation, on his match labels in the 1870s. His business closed and the trade mark was taken over by Moorelands, who relaunched England's Glory (**3**) matches in 1891. Many ingenious promotions made it popular, including competitions for the best models made from empty boxes.

3

1

2 | 4

It was a Frenchman, La Croix, who, in 1796, first manufactured a fine cigarette paper in which to roll tobacco. The Rizla (**2**) pack here dates from about 1890. The brand name was developed from riz – the French for rice – as a paper similar to rice paper had been used, and the first part of La Croix; the latter part of the name, Croix, means a cross, the emblem displayed after the brand name, Rizla.

Swan matches were originally manufactured in 1883 by the British firm of Collard & Co of Liverpool, who were taken over in 1895 by The Diamond Match Co, an American firm which relaunched the matches as Swan White Pine Vestas in 1897. Diamond Match then merged with the London match firm Bryant & May in 1901. The brand was shortened to Swan Vestas (**4**) in 1906. The Swan changed direction in 1959.

FLOR DE

FERNANDEZ GARCIA

HABANA.

NEPTUNO 170 & 172.

CASA FUNDADA EN 1876.

5

Some of the finest quality commercial printing was that done for cigar labels at the end of the nineteenth century. The Cuban cigars frequently won medals, *which were used to embellish the already decorative labels. The label for Fernandez Garcia (5) has gold medals embossed out with the rest of the pictorial details.*

SPIRITS

1

2

One of the larger whisky distillers in Ireland was Dunville & Co of Belfast, makers of Dunville's VR Whisky (*1*). The label design may have pre-dated the 1880s (it was still used in 1920), and contained many shamrocks but also the English rose and Scottish thistle.

Jack Daniel, who came from Tennessee, bought a distillery at the age of 17 in 1865 in the next year, he rebuilt it and in 1887 Jack Daniel's Old No 7 Tennessee Whiskey was born. The square bottle was adopted in 1895.

The recipe for Southern Comfort Liqueur Spirit was the secret of M. W. Heron, a bartender in New Orleans in 1875. The taste for this liquor grew and by the 1890s it was being bottled from his bar at Memphis. Heron moved up the Mississippi River to St. Louis in the early 1900s.

Both bottles here (*2*) are modern. The Jack Daniel's label still retains its original lines, while the label for Southern Comfort was redesigned in 1945.

The recipe for White Horse Cellar Whisky originated in 1746. In 1890 the distiller, Mackie & Co of Glasgow adopted the white horse symbol (**3**), which was taken from the White Horse Cellar Inn in Edinburgh.

The French distillers, Eugène Vincent & Co of Lyon, had been established since 1807. They made a wide range of drinks – from brandy and liqueurs to rum and kirsch. They came in all manner of shapes and sizes of bottle. The two shown here are from the 1890s. Glacial Peppermint (**4**) had an additional label to show off the medal won in 1889, and Marasquin (**5**) came in a protective layer of raffia.

3

4

5

FORTIFIED WINES

1

The fortified wine known as port comes from Portugal. In the mid-eighteenth century bottles of port started to be identified with simple labels. By the end of the nineteenth century they had developed into highly colourful and decorative designs, such as those for Condessa, Puritano, Gloria do Porto and Burguezia (**1**).

2

Vermouth, a white wine flavoured with aromatic herbs, was served as an appetizer in the nineteenth century. In Italy a list was issued in 1840 permitting only certain firms to produce vermouth. One of these was run by Italians Martini and Sola in the wine-growing district of Turin. In 1863 Rossi joined the firm and the first bottling of Vermouth began. The name was changed to Martini & Rossi in 1879. It was probably as early as the 1860s that the well-known label design with the factory, flags and women had begun to take shape. By the 1880s the design was at its most complex (*2*), including gold medals won at exhibitions. The Martini & Rossi label was often imitated although some labels were totally different, such as that for Vermouth Suisse.

BEVERAGES

Runkel Brothers (**1**) of New York founded their business in 1870. During the 1880s and 1890s most of their "pure breakfast cocoa" would have been distributed from bulk containers. Solid tins with colourful sides and airtight screw-on lids could be purchased with half-a-pound of cocoa, made from the "choicest cocoa beans". On this tin, c 1890, a royal coat of arms was displayed to give the impression of royal approval.

1

The Great American Tea Company imported Chop Tea (**3**) from China. The airtight tins in use during the 1890s helped keep the contents fresher for longer periods. In China a trade mark was known as a "chop", which here was a double sun.

The elaborate tin for Super Café Camara (**4**) came from Brazil in the late 1890s. Their trade mark was a lady holding a cup of coffee sitting on a globe.

3

2

James Horlick, the British creator of an artificial infant food called Horlick's Food, emigrated to America in 1873 to join his brother William. There, they formed a partnership to produce the food, reformulating it in 1883 as Horlick's Malted Milk (**2**). James returned to England around 1890 to establish a London branch, followed by a manufacturing unit in Slough from 1906.

The Horlick's outer wrapper here has been converted into a dummy British display carton, but the basic design dates from the 1880s.

4

John Cadbury started his career in 1824 as a tea dealer and coffee roaster in Birmingham, England, and began manufacturing cocoa and chocolate from 1831. In 1866 Cadbury's became the first British company to produce a pure, concentrated cocoa, known as Cadbury's Cocoa Essence (**5**). The tin here dates from the 1880s. On the side of this 42oz tin an extract from The Lancet states, "We have examined the samples brought under our notice, and find that they are genuine ... the majority met with in the shops are not genuine, but consist of mixtures of cocoa, sugar, and starch".

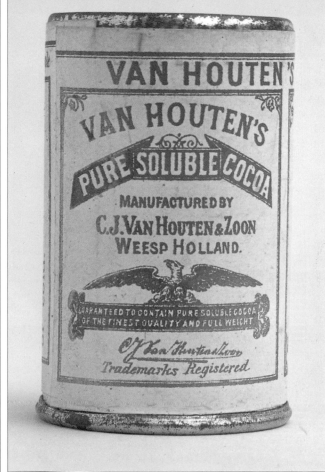

6

5

As far back as 1828 a Dutchman called Van Houten had formulated a pure cocoa. By the end of the nineteenth century, this had been developed into a spicier taste and was being exported during the 1890s with a pure white label with gold lettering. The tin shown here (**6**) is a miniature sample only 4.5 cm (2 in) tall.

BREAKFAST CEREALS

Most of today's breakfast cereals were first introduced in America, where supplies of cereal grain were plentiful and there was a growing public awareness of health foods. The leading company in 1890 was the American Cereal Co whose Granulated Hominy (**1**) was considered "very healthy and palatable in any form, boiled as a mush for breakfast, or served as a vegetable for dinner, fried in fritters or griddle cakes".

Grape-Nuts (**2**), a food for "brain and nerve centres", was the creation of C. W. Post in 1897, who had set up business in Battle Creek, Michigan, two years earlier.

1

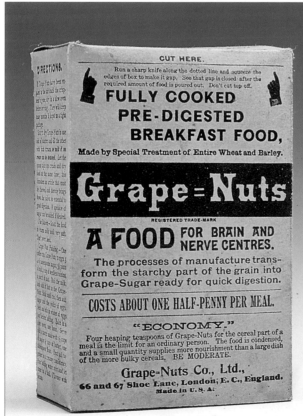

2

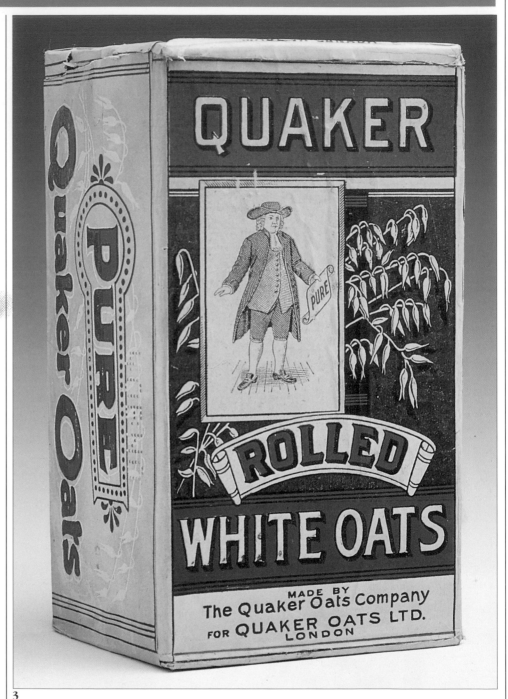

3

Quaker Oats (**3**) were first marketed in 1877 by a small American firm, the Quaker Mill Co of Ohio, and the Quaker emblem was used from that time. After a series of changes in ownership, Quaker Oats were in the hands of the American Cereal Co in 1890. It became their biggest selling brand, and around 1900 the firm's name was changed to the Quaker Oats Co.

SPICES

The design on the Huyler's Cocoa tin (**4**) is typical of American style in the 1880s. The wholesomeness of the product has been accentuated by depicting the cocoa pods hanging from the branches, and the message on the front was "purity and deliciousness of flavor unexcelled".

The French spice tins, Epices Rabelais (**5**), show a gruesome scene of a butcher about to carve up a pig. Amongst the flowers and foliage design of the 1890s, a panel illustrates the fourteen medals awarded to the firm.

4

5

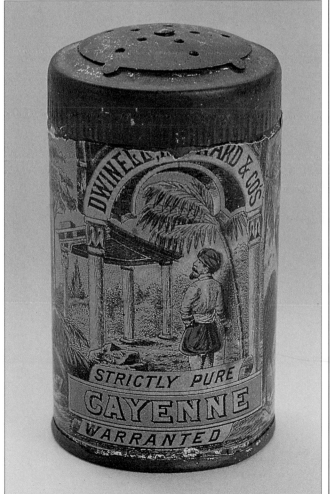

6

7

For Dwinell, Maynard & Co's Cayenne (**6**) of the 1890s, the pictorial theme was an Eastern setting, and the message for the American public was an assurance that the spice was guaranteed "strictly pure".

The English agents Walter Preston Ltd for American Kamwood Extract (**7**) continued to use the American 1890s design on their tins. Down the side of the tin was written, "This is the original. All other colourings are vile imitations". Walter Preston Ltd of Leeds also produced Yorkshire Polony Seasoning, antiseptic rusks and all classes of food preservatives.

BISCUITS AND CRACKERS

1

It was in 1857 that James Peek, a retired tea merchant, joined forces with George Frean, a West Country miller. Together they formed Peek, Frean & Co and set up a factory in London. After 1870 the firm concentrated on fancy biscuits. The biscuit tin depicted here (*1*) is covered by a paper wrapper probably designed in the late 1890s, and still in use forty years later.

Robert McVitie had a baking business in Edinburgh, Scotland, in the 1840s. By the 1880s his biscuits and cakes were winning international medals. In 1888 Charles Price, a former travelling salesman with Cadbury's, joined McVitie's and such was his prowess at selling that he was made a partner in the firm.

The McVitie & Price label (*3*) for Royal Scot biscuits would have been designed in the mid-1890s.

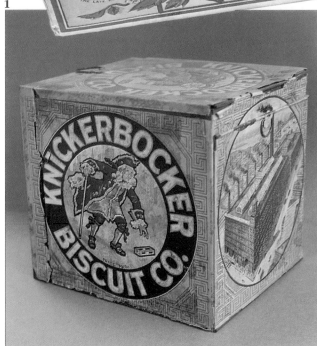

2

The large outer tin for Knickerbocker Biscuits (*2*) would have contained 10lbs of cookies. On one side of this 1890s tin a lable proudly showed the great American factory that produced the contents.

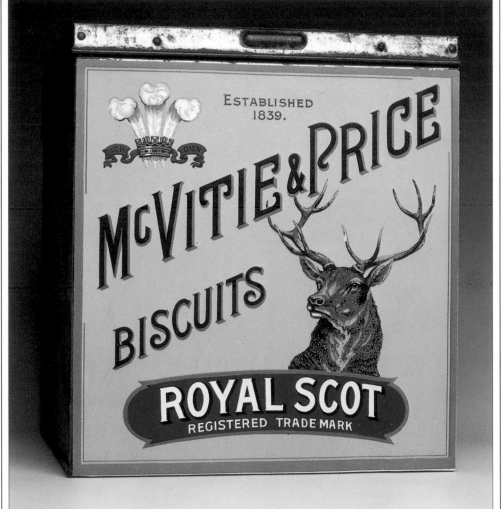

3

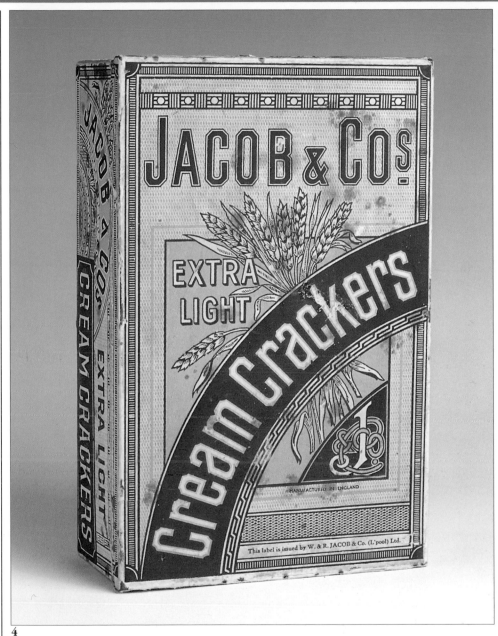

4

5

6

In Ireland during the eighteenth century, the Jacob family started to bake bread and ship's biscuits. In 1851 they decided to diversify into fancy biscuits and, as business grew, they moved premises to Dublin. Although the 1860s saw financial difficulties, by 1880 business was flourishing once more. George Jacob visited America in 1884 and witnessed the success of crackers there. The following year he launched Jacob's Cream Crackers (4) in Britain. This pack design was still in use in the 1930s and continued to be used thereafter for exports.

In 1846 the family of Lefèvre-Utile set up as pastry-cooks in Nantes, France. They soon built up a reputation for their biscuits. The son, Louis Lefèvre-Utile took over in 1883 and extended the family business into an industrial firm. After a visit to England, whose biscuit manufacturers took the vast majority of European sales, Louis himself launched into the manufacture of dry biscuits. One of his great successes was the creation of Petit-Beurre (5). This pack design dates from around the mid-1890s, although after Petit-Beurre won the Grand Prix at the Paris Exhibition of 1900, appropriate additions were made to the pack.

The biscuit firm of Pernot from Dijon, France, followed the traditional style of biscuit labels by using a double circle design. In this case (6), late 1890s, the central space is filled by a soldier who rests to eat a biscuit.

CANNED FOODS

It was in France in the early nineteenth century that the first experiments in canning took place. Nicolas Appert began preserving meat, vegetables and fruits in glass jars, carefully testing how long each food needed to be heated. In 1810 he published his findings on preserving food in airtight containers. He realized, for instance, that vegetables "must be picked as freshly as possible, and treated with the greatest speed, from the garden to the cooker in a single bound". Fifty years later the French chemist, Louis Pasteur, discovered why airtight containers were necessary – to avoid the growth of bacteria.

Once a mechanical means of manufacturing tin cans had been found, during the 1850s, canned foods became more readily available. At first they were used largely by explorers and the military, and then gradually the public for use away from home.

The canned peas, meat paste and truffles shown here (1) come from France in the 1890s. The medals awarded to Lebreton & Bree of Paris go back to 1867.

2

In America glass jars were initially used for preserving foods, but by the 1840s tin cans had become more widely available. At many of the sea ports, fisheries had their own canning plants for salmon, crabs and oysters, and by the 1850s all kinds of fruit and

vegetables were being canned.

It was probably not until the 1860s that colourful labels started to appear on the cans, but by the end of the century there was an extraordinary range and variety of American can labels.

The labels here (**2**) date from the 1890s, and come from different parts of the United States – Defy The World Tomatoes packed by Kelty & Son of New Jersey, Yellow Free Peaches from the San Francisco Fruit Canning Co, Butterfly Brand Boston Marrow

Squash from Olney & Floyd of New York State, California Fruits from San Jose Fruit Packing Co of California. The Seasons Brand Salmon label was possibly of British design, produced as a stock label for cans of salmon exported to Britain.

CANNED FOODS

1

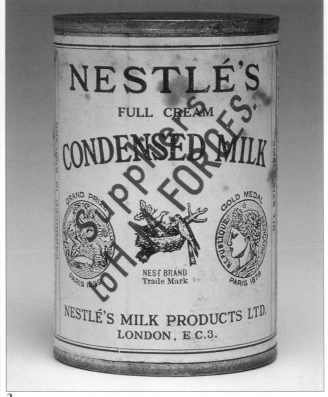

2

In 1868 the firm of Libby, McNeil & Libby was set up in Chicago by two brothers and a friend to produce corned beef. To start with barrels were used, but in 1872 some corned beef started to be packed in tapered tin cans. The design for this label (**1**) probably originated in the 1870s, being modified to show the larger premises taken over in 1887.

Henri Nestlé of Vevey, Switzerland, first produced condensed milk in 1875. By the 1890s the Nestlé's Condensed Milk (**2**) had been developed, with a label similar to that of the 1930s one shown here. Up until the 1930s, the cans were also individually overwrapped.

Petits Pois (**3**) were made by Gontier Frères in the late 1890s. The design has been printed directly onto the can.

Many of the early can labels did not go right round the can. In the case of the Pure Honey (**4**) produced by N. Liljedahl of California, the label covers only half the side and does not reach right down to the can's base. The label dates from about 1885, and it is interesting to note that there is no brand name, nor is the company's name given much prominence.

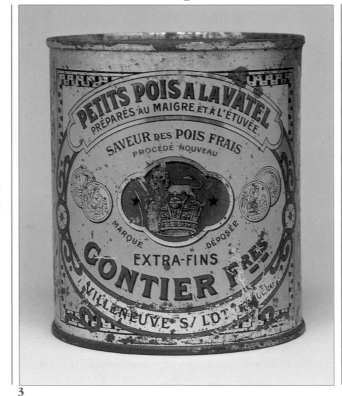

3

4

CONFECTIONERY

5

The firm of Van Haaren-Nieuwerkerk, established in 1770 in the Hague, Holland, first produced coffee-flavoured bonbons in 1794. They were the idea of the Austrian Ambassador to the Netherlands, Baron Hop, hence the name Hopjes given to these delicacies. The picture of children outside a sweet shop dates from the 1890s (5).

6

7

8

The varieties of Turkish Delight made by Lipton (7) and A. P. Frangos (6) ("imitated, but not equalled") both use an Eastern design with an identical border, and date from the late 1890s.

Traditionally dates have been packed in boxes with a wooden top and base and cardboard sides, covered with coloured paper and with a decorative label. Occasionally a manufacturer issued a special tin to make the product worthy of presenting as a gift. This tin for Champagne Frères' Tunis Dates (8) is from the late 1890s; a picture of their standard date box served as confirmation that the dates were from the same source.

HEINZ

1

2

3

Henry J. Heinz formed a partnership with L.C. Nobel in 1869 bottling Celery Sauce (2), Sweet Pickles (1), horseradish and so on. After a few years of success the business ran into problems, so Heinz went to work with his brother and a cousin in a preserving firm. By 1888 he had paid off his debts. The three emblems that distinguished Heinz products were the keystone shape used from 1880, the pickle mark devised in 1889, and in 1896 the slogan "57 Varieties" as on the Tomato Soup can (3) which seemed to H.J. Heinz to be the right sounding number.

4

5

Henry Heinz made his first trip to London in 1886 when he sold products to Fortnum & Mason. A factory was established in London in 1905. The Baked Beans (4) in cans with paper overwraps of the late 1890s and Apple Butter (5) in a stone jar with handle date from about 1895, and show the everwidening range of Heinz products.

CAMEMBERT

French Camembert cheese labels traditionally depicted rural scenes of dairy maids *(2&3)*, cows in fields and the model farms *(1&4)* where the cheese was manufactured. Occasionally a label would be given a more robust image such as that of Napoleon. The designs were probably created in the 1890s although they were updated to include, for example, the motor vehicles which replaced horse-drawn carts.

1

3

4

DRIED GOODS

Invented by two American pharmacists in 1863, Royal Baking Powder was the first raising agent that could be used without the addition of acid such as is found in sour milk. The chemists moved in 1875 to larger premises in Chicago. The Royal Baking Powder Company was set up after the move to New York.

The basic design (**1**), which still survives today, dates from around 1890. The pack shown here has the words "cream of tartar" added, along with the new manufacturers' name, Standard Brands Inc. Royal was one of the founding companies that combined in 1929 to form this new conglomerate.

1

2

One of the great French delicacies, pâté de foie gras (**2**), was traditionally packed in substantial earthenware pots which helped to keep the contents cool. Centred on Strasbourg, two of the leading firms were J. G. Hummel (founded in 1827) and Louis Henry (founded in 1829).

Sea Moss Farine (**3**) was an early form of instant pudding that, with liquid added, made a blancmange. It was a product with sales in both America and Britain.

3

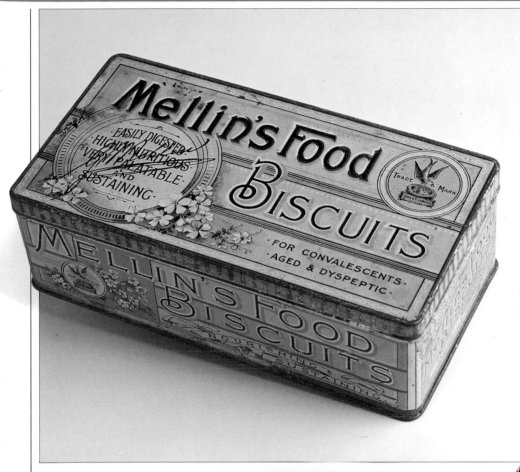

For many babies born between the 1880s and 1930s, nourishment was derived from Mellin's Food of London, which came either in the form of a powder mixed with milk or as biscuits. Mellin's Food Biscuits (**4**) were also recommended for athletes, travellers, dyspeptics and invalids. The company's outlets were established around the world, including North America, India, Australia and New Zealand.

4

Most American housewives baked their own cakes at home, but at the end of the nineteenth century, cakes were also made by the many local bakeries. Occasionally a slice of rich fruit cake was packed in a tin to make it presentable as a gift, as with this Kennedy's Celebrated Fruit Cake (**5**).

5

6

The invention of the chemist George Weddell, Cerebos Salt (**6**) first appeared in 1894 and was manufactured at Newcastle upon Tyne, England. It contained phosphates that strengthened bones and teeth. The tin later became known for its "patent pourer" enabling the salt to be easily dispensed. In 1906 Cerebos took over Birdcatcher Salt whose trade mark was the boy chasing a chicken. In 1919 this image first appeared on the Cerebos tin.

SOAP AND POLISH

Many successful household products have been sold on both sides of the Atlantic, such as the brands illustrated here. Rising Sun Stove Polish (3), manufactured by J. L. Prescott Co. of New York, was claimed to be absolutely safe. "No housekeeper can afford to take chances of being painfully burned and perhaps horribly disfigured for life by using inferior, inflammable and explosive stove polishes put up by inexperienced or unscrupulous manufacturers." This brand was sufficiently successful for the British company Reckitt & Sons to buy the rights and use the same trade mark (1).

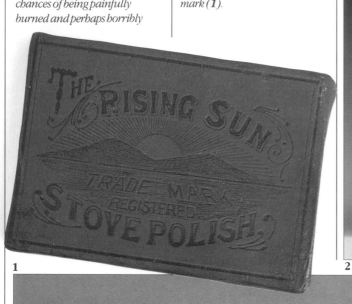

1

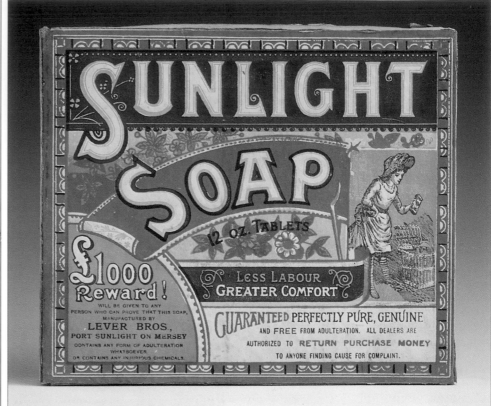

2

3

The name Sunlight, registered by William Lever in 1884, was used for the best soap that was being made for him. The following year he started to produce his own soap from a factory at Warrington, England. This was the first soap to be sold in a carton; it had previously been wrapped in parchment papers. In 1889 Lever's factory moved to a site on the banks of the Mersey, later renamed Port Sunlight. By the mid-1890s the colourful Sunlight pack (2) with its series of different illustrated scenes had become known around the world – in America, Australia, China and South Africa, and throughout Europe.

*Like Rising Sun, Enameline (**4**) was another product of J. L. Prescott that was introduced to Britain by Reckitt & Sons in the late 1890s. The same overall design was used in both countries.*

4

*Monkey Brand (**5**), a scouring soap for cleaning pots and pans, was first manufactured by Benjamin Brooke of Philadelphia. In 1899 Lever bought the company and the brand was introduced to Britain under the slogan "won't wash clothes". As with many such advertisements at the time, fanciful claims were made: "makes tin like silver, copper like gold, brass ware like mirrors". A powder form of Monkey followed, soon to be overtaken by Lever's own brand, Vim, launched in 1904.*

5

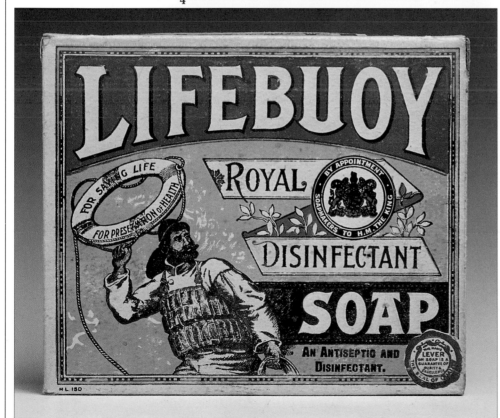

6

*Lifebuoy Soap (**6**), "for saving life and preservation of health", was launched by Lever Brothers in 1894. This household soap contained carbolic which gave it disinfectant properties. Sales campaigns often followed in areas where there had been an outbreak of disease. Although not immediately as successful as Sunlight, its introduction encouraged many other firms to start producing their own carbolic soap.*

MEDICINAL

Effervescent Limosine health salts (*1*), prepared from ripe lemons, limes and grapes, were manufactured by Oppenheimer Brothers of London. Primarily for export, the box was typical of 1880s' style and must have cheered the patient.

Thomas Beecham (*2*) had been producing his famous pills since the 1850s. The business expanded in the 1880s when the product was nationally advertised and described as being worth a guinea a box.

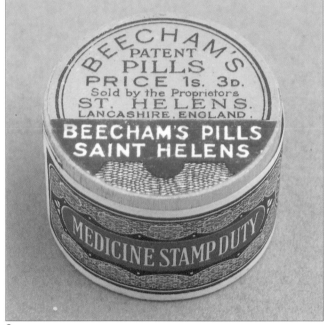

1   2

3

Pastilles Valda (*3*), for quick relief of all bronchial affections, were manufactured in Paris

and London. The design is typical of the confusion of styles prevalent in the mid-1880s.

4

Diamond Brand Pennyroyal Pills (*4*) were manufactured during the 1880s by the Chichester Chemical Company

of Philadelphia. The firm was anxious to confirm that these pills for relieving menstrual problems were "the only reliable female pill

now offered for sale", and attention was drawn to the trade mark of a red cross in the middle of a diamond.

Eno's Fruit Salts (**5**) were first produced in Newcastle upon Tyne in the 1860s, and found a ready market with the seamen of the port. In 1880 J. C. Eno moved to London and started to export his fruit salts all over the world. It must have been soon after that the colourful outer wrapper was added to the package. Not only did it provide protection to the glass bottle, but gave it an identifiable image that distinguished it from the hundreds of other medicinal products. Eno's were purported to cure digestive ailments, disorders of the liver, feverish and rheumatic conditions.

6

5

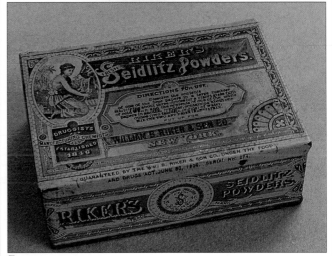

7

Typical of French medicinal packs in the 1880s and 1890s was that of Réglisse Florent (**6**). As with early pill and potion packs in Britain, there appeared to be little need for colour or pictorial image, except when a royal coat of arms was awarded.

In the USA, medicinal products were given a more elaborate design. This is true of the Seidlitz Powders (**7**) produced by Rikers, manufacturing chemists in New York since 1846. Indeed, the design style was more akin to that used on the cocoa and coffee tins of the time.

PERFUMERY

In 1834 Eugene Rimmel joined his father to set up a perfume business in London. He published The Book of Perfumes in 1864. Florida Water (1) was just one of a wide range of toiletries that he sold.

Aqua de Florida (2) was made by Murray & Lanman of New York from about 1880. The distributors were Lanman & Kemp, who exported all over the world including South America where this bottle comes from. The label imitated that of Rimmel (1) and is still in use today.

1

2

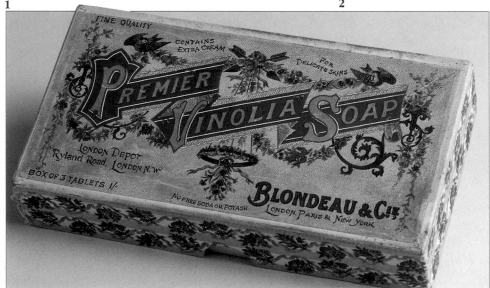

3

In 1885 the firm of Blondeau et Cie was founded in London to manufacture toiletries. One of their most successful products was Premier Vinolia Soap (3). By the end of the nineteenth century the firm had adopted the brand name Vinolia as their company title.

4

Scented sachets were once very popular as women liked to carry them next to their handkerchiefs or lay them amongst their clothes in wardrobes and drawers. The fragrance the sachets gave out was usually the same as that of the perfumes available in bottles. The Spanish sachet, Polvos de Arroz (**4**), dates from the late 1890s; Okell's Scent Sachet (**6**) with the nightingale from about 1895. and the Delvoix Exquisite Perfume Sachet (**5**) from the 1880s.

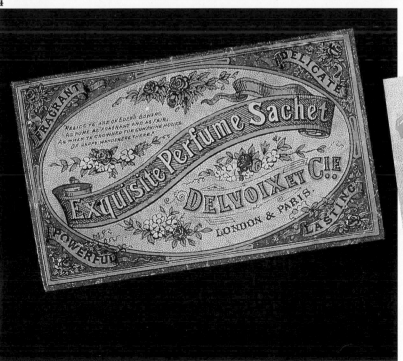

5

6

Manufactured in France by the
Vichy Société Centrale around
1905, this tin of medicinal
sugar pastilles was highly
decorated on both front and
back. The pastilles were made
from the mineral salts extracted
from the Vichy water, and were
said to improve digestion.

CHAPTER TWO
NOUVEAU INSPIRATION

1900-1919

INTRODUCTION

By the turn of the century, the grocer's store had become more of a centre for the distribution of pre-wrapped branded goods than a traditional shop where the provisions could be inspected before purchase. The customer was now having to rely more on the word of the manufacturer, communicated by an impersonal message on a poster or on the pack itself, than on the grocer's recommendation. Indeed the skills of the retailer were fast becoming redundant.

Everything from jam to breakfast cereal, from household soap to scouring powder, was being individually packed by the firms at their factories. In Britain, a great battle raged between the tea companies with their branded products and the grocers who blended loose teas to their customers' own requirements.

For some time, grocers had resisted the tea brands which were delivered already packaged; these products were, after all, making their craft as tea blenders extinct. Also (for the more unscrupulous) it prevented the addition of "make-weight" to the tea. The manufacturers found that it was the chemist and not the grocer who more willingly retailed the pre-wrapped tea. Not only were chemists used to selling wrapped articles, but teas such as Ty-phoo were being promoted as good for the digestion – "doctors recommend it". Furthermore, some tea companies were making much of the fact that the weight

The French cosmetic firm of L. T. Piver was founded in 1774, and based in Paris. Medals had been won at many of the early international expositions such as the one held at London in 1851 and another at Paris in 1855. By the end of the 1860s the company was producing over 800 packaged varieties or sizes of perfume, cologne, toilet soap, pomade, brillantine, dentifrice, powders and beautifiers. The range of cosmetics under the Pompeia brand name (1) arrived in 1907 at a time when this classic style was in vogue.

1

2

messages on the pack. Medals won at exhibitions, guarantees, testimonials and royal crests all helped to give credence to the product, but failing those (or in addition to them) inducements could be offered.

It was during the early 1900s that manufacturers made the "premium gift" king of promotions. On everything from tea and meat extract to household soap and tobacco the offer of a free gift induced the would-be customer to buy a particular brand. By keeping part of the pack the customer could save up for all manner of articles. In 1902 Ogdens were the first company to issue coupons in cigarette and tobacco packets. The coupons could be redeemed for merchandise selected from their "list of presents". In 1904 purchasers of Quaker Oats were able to obtain silverware, handkerchiefs, stockings, socks and pipes at less than half the retail price if they sent off the required number of trademark emblems. At the same time cereal bowls were given away by the company; 35 packet emblems secured four bowls, and proved a popular promotion.

Another idea was devised by British confectionery firms who issued tokens in their cocoa tins. Once enough had been saved they could be sent off in exchange for a special decorative tin filled with chocolates.

For the competing cigarette companies a promotional success was scored with the issuing of pictorial cards in each cigarette packet. The idea had been used first in America and then in Britain during the 1880s. The cards had the dual purpose of protecting the cigarettes in their

First used as a brand name in 1892, Del Monte was part of the Oakland Preserving Company which in turn became part of the California Packing Corporation in 1916. Most of their products were canned fruits and vegetables, but other foods such as condensed milk (2) were also produced.

The National Biscuit Company introduced Barnum's Animal Crackers (3) in 1902. Animal shapes had been used before by several American bakeries, but the original idea had come from England.

indicated on their ¼lb packs referred to the tea alone, and did not include the wrapper.

In America, the National Biscuit Company had begun to have some success with its In-er-seal carton. By the turn of the century, biscuits in this new type of hygienic container were becoming as popular as those sold loose from the traditional large glass-fronted tins and wooden boxes. Manufacturers also found a good market for the novelty biscuit. One such line was Barnum's Animal Crackers, named after the famous circus owner and sold in a box that looked like an animal cage.

World War I further encouraged individual packaging, since it was much easier to distribute and supply rations to troops in the field if they were packed in handy sizes. So long as a can opener was on hand, a tin of bully beef could make a meal ready in a moment. The convenience of such products became obvious under the uncertain conditions of war-time Europe. New convenience foods like Oxo Cubes were much in demand and thrived – one cube in a billy-can of hot water provided immediate sustenance.

As foodstuffs were packed increasingly in individual tins, cardboard boxes, paper wrappers and glass containers, there was a new problem for the manufacturer. If the cocoa was already sealed in a tin, it could no longer be inspected before purchase. Its quality, texture, aroma and freshness all had to be accepted by the customer; its quality had, in the first instance, to be assured by the image and reputation of the manufacturer. The ultimate test was in the tasting, but even prior to this a certain amount of influence could be brought to bear by the judicious use of

3

INTRODUCTION

1

Yellow Kid Ginger Wafers (1) produced by Brinckerhoff & Co of New York, c 1900, was one of many brands to use this popular character. The Yellow Kid had been created by Richard Outcault for a colour comic strip that first appeared in the New York World *in 1895. The following year the Yellow Kid also appeared in a rival paper, the* New York Journal, *and Outcault began to license the Yellow Kid for a variety of products.*

During the early 1900s Art Nouveau had considerable influence on package design. Most designs were "locked" into those that had been created during the previous thirty years or so and it would have been a foolhardy firm that changed a well-established product image. But brands established at this time were able to take advantage of the style of the moment. The ever-volatile cosmetic, perfume and toiletry markets (especially in France) revelled in the flamboyancy of Art Nouveau, and today the packs seem just as alluring as when they were first produced. Other packaged products, such as Kellogg's Toasted Corn Flakes (launched in 1898), absorbed the style in a more subdued way.

2

flimsy packs and serving as a major inducement to buy a particular brand. As cigarette smoking grew more popular, so too did the collecting of various series of cigarette cards. On the outbreak of World War I in 1914 the manufacturers in Britain began issuing cards depicting war heroes, war maps and stand-up cut-outs of soldiers and their uniforms. (Shortage of paper in World War II and agreement not to continue their production ended this form of promotion.)

A different way of attracting attention was to add a popular character to the pack. This had happened in 1896 when the Yellow Kid, a character from a comic strip in the *New York World*, had been used on the wrappers of products like biscuits, candy and cigars. The creator of the Yellow Kid, Richard Outcault, then went on to market his next character, Buster Brown, who had appeared in the *New York Herald* from 1902. In the same year, the makers of Force cereal introduced their own character salesman, Sunny Jim. Ever since, character merchandizing has been part of the promotional scene, from Mickey Mouse to the Ghostbusters, from the Green Giant to Superman.

Whereas the impetus for marketing innovation had come from Britain and the United States, it was in France that the most dramatic changes in design style took place, during the early 1900s. The movement, known as Art Nouveau, had grown up in France during the 1890s, a style characterized by sensuous lines, and often depicting beautiful women with flowing hair and intertwining flower motifs coloured in muted pastels. The prime exponent of the movement was Alphonse Mucha, a Hungarian living in Paris.

Force Wheat Flakes were introduced in 1901 by the Force Food Company of Buffalo, New York State. They were the first breakfast cereal to be imported to Britain, and in the year of its

introduction, 1902, the character used to promote Force was Miss Prim. She was replaced in the following year by Sunny Jim. This British pack design (2) dates from the late 1910s.

On the technical front there were continuous developments in packaging materials. New ways of opening a package (such as the cord serrator), advances in dispensing (such as the sprinkler tops for talcum powder tins), and better ways of closing and re-sealing were all part of the endless search for better packaging techniques. New improvements were exhibited at the prestigious world fairs, such as the Universal Exhibition in Paris of 1900. It was on this occasion that a Frenchman, Louis Chambon, displayed a carton-printing machine that was able to print, fold, fill and seal a carton in one continuous operation. Such a machine further encouraged firms to supply their wares to the retailer in a pre-packaged form.

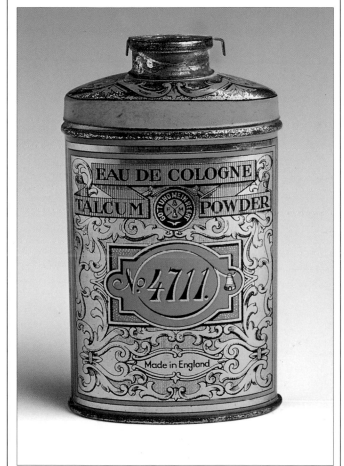

3

4

The basis for the design of this 4711 talcum powder tin (3) would appear to go back to the 1870s, but the flattened oval shape powder tin came in after 1909. The company was founded by a German banker, Ferdinand

Mulens, from Cologne, at the beginning of the nineteenth century. He had been given a secret formula for eau de cologne by a monk; the basis of the formula involved the number 4711.

It was during this period, although having little effect at the time, that two major new materials were first introduced to the packaging repertoire. By 1910 both America and Britain were producing aluminium foil: then in 1912 Cellophane film, a derivative of plastic, was invented by a Swiss chemist, and manufactured shortly afterwards in France.

By the end of this period, a great abundance of brands had been established offering new culinary delights as well as a range of household washing powders that purported to increase the efficiency of the weekly wash. Many of the new food products were those that served as instant nourishment, needing little preparation – Kellogg's Corn Flakes, "the 30-second breakfast", is one of the best known of the survivors.

This wrapper for an individual tablet of toilet soap, Savon Idea (4), was created by the prime mover of the Art Nouveau movement, Alphonse Mucha, c1900.

BREAKFAST CEREALS

1

A lawyer, Henry Perky, created Shredded Wheat by accident. Production started in 1893. In 1901 a large factory was built near Niagara Falls, and the packets were redesigned (**1**) to show off Perky's new "palace". Nabisco took over in 1928.

Quaker Puffed Rice (**3**) was patented in 1911. The cereal was based on a process that swelled rice grains to eight times their normal size. It was recommended as being delicious served with any fresh or cooked fruits.

3

Dr John Kellogg was a supporter of healthy eating and experimented enthusiastically with health foods. In 1894 his first dried wheat flakes were produced. His younger brother, Will joined him in 1898 and in 1906 W. K. Kellogg started to sell Kellogg's Toasted Corn Flakes (**2**) – "None genuine without this signature". The British company was set up in 1924. A new cereal, Rice Krispies, was launched in 1928.

2

Mother's Oats (*4*) were made by the Great Western Cereal Company of Iowa, but in 1911 they were bought out by their rivals, Quaker Oats. The pack here retains much of the original image – the dress has been modified to look less dated, and some lettering has been removed.

Post Toasties (*5*) were launched in 1915 by the Postum Cereal Company, Michigan. They were promoted as being "double-thick" and were said to be "expertly seasoned and toasted by special process to golden brown". They also "hold their crispness and their flavor, even when swimming in milk or cream".

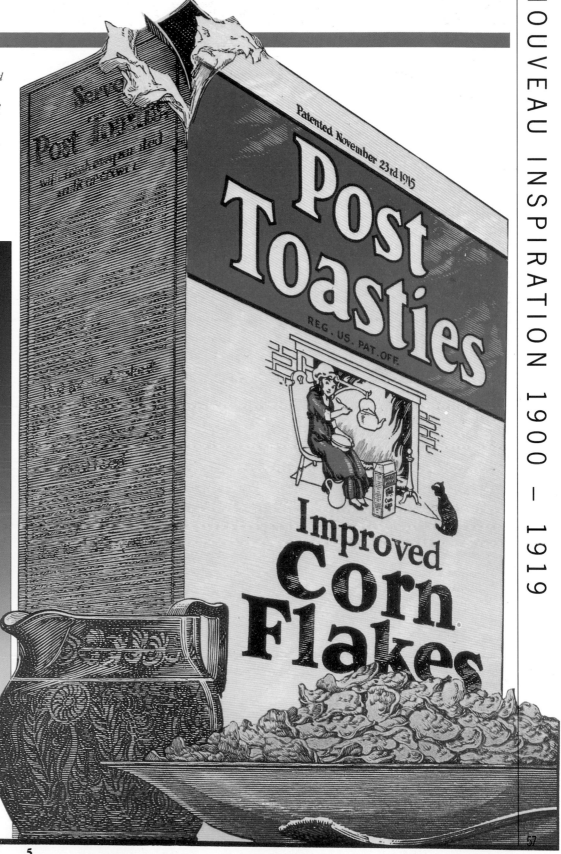

CONFECTIONERY

Rademaker's Hopjes (**1**) had been made in Holland since 1875. This tin dates from about 1910.

The firm of Tucker & Son made their Devonshire Cream Toffees (**2**) in Totnes, England. The label design with its meadow

flowers trade mark was probably created around 1900.

1

2

3

From the USA in about 1910, Mellomints (**3**) came in small tins that were colour-coded according to the flavour inside; for example, green was used for lime-flavoured sweets.

Cadbury's Dairy Milk Chocolate was launched in 1905, followed three years later by Cadbury's Neapolitan (**4**). A plain chocolate bar, Bournville Chocolate, arrived in 1910, in a contrasting bright red pack with sweeping gold lettering.

4

The French chocolate industry was well established by the end of the nineteenth century with many firms producing similar bars. The attraction of the wrapper was crucial in securing sales, and by 1905 chocolate wrappers (**5**) vied with each other in a riot of colour, illustration and lettering.

Americans were chewing gum in the 1850s and in 1892 William Wrigley first sold gum in Chicago; next year he launched the Spearmint and Juicy Fruit brands. Beech-Nut was founded by Bartlett Arkell in 1891. Beech-Nut Chewing Gum arrived in 1911, about the date of these packs (**6**).

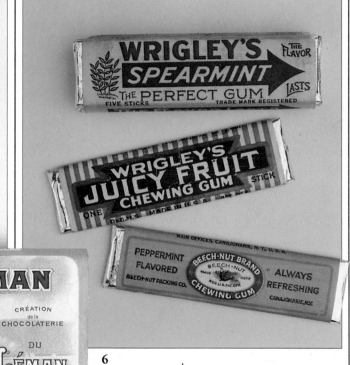

6

5

Since 1868 Jean Tobler had sold chocolate specialities from his shop in Bern, Switzerland. They did so well that he set up a factory and by the end of the century he was exporting to Britain. It was at this time that Toblerone (**7**) was first produced. Made of milk chocolate with almonds and honey, the distinctive mountain range shape of the bar is probably the only chocolate mould ever patented. Incidentally, the signature on the patent form was Albert Einstein's, who was working in the Swiss Patent Office at that time.

7

BISCUITS

A biscuit can be an everyday food but once packed in a fancy box with a decorative design it immediately becomes an attractive gift. Thus Butter Thins (*1*) made by the National Biscuit Company and packed in a tin decorated with a design of buttercups take on the mantle of a presentable gift (c 1915).

The two biscuit boxes from Lefèvre-Utile were designed for their appeal. The Gaufrettes Pralinées (*2*) label was created by Alphonse Mucha around 1900. The Miarka Gaufrettes (*3*) label probably dates from a little later; unfortunately the artist did not sign it. Thanks to the imagination of Louis Lefèvre-Utile good artists were employed, in the same way that they were for the posters which advertised the biscuits.

1

2

3

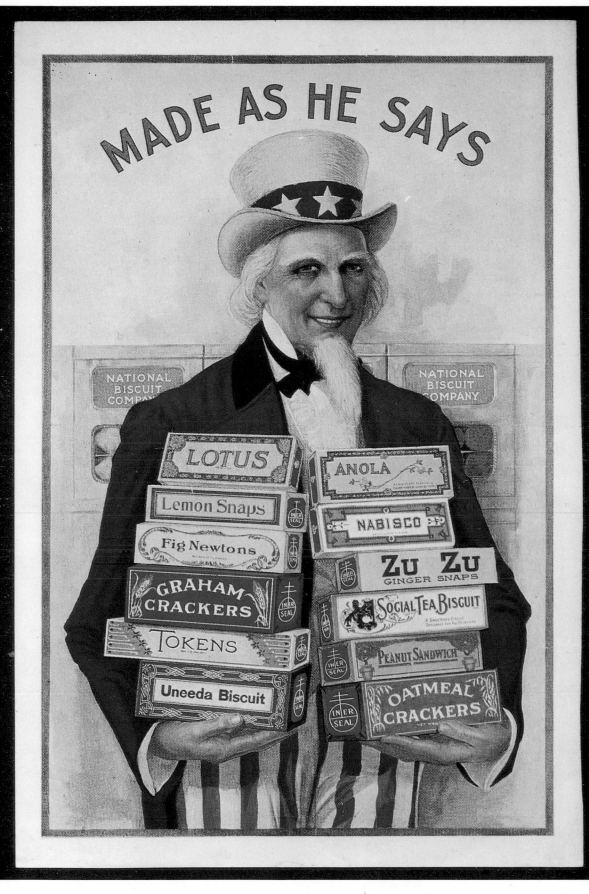

MADE AS HE SAYS

LOTUS

Lemon Snaps

Fig Newtons

GRAHAM CRACKERS

TOKENS

Uneeda Biscuit

ANOLA

NABISCO

ZU ZU GINGER SNAPS

SOCIAL TEA BISCUIT

PEANUT SANDWICH

OATMEAL CRACKERS

When the National Biscuit Company – an amalgamation of three large bakery firms and numerous smaller ones – was formed in 1898, the list of products manufactured ran into several thousands. During the early 1900s there was a period of rationalization when many lines were discontinued, especially in bread and confectionery. By 1908 there were only a few hundred biscuit, cookie and cracker lines, including 44 products sold in the In-er-seal cartons.

*In the patriotic advertisement here (**4**) of 1918, Uncle Sam holds a dozen popular lines. The Peanut Sandwich had just been introduced, but most of the others had been on the market for over ten years – Fig Newtons had been made since 1892, Uneeda Biscuit since 1899 and Nabisco Sugar Wafers since 1902.*

FACE AND FEET

Early in the nineteenth century, William Yardley, a manufacturer of swords, buckles and spurs, took over the ailing soap business of his son-in-law, William Cleaver – a business started by Cleaver's father in 1770. Around 1900, the Yardley name began to be promoted and a range of perfumes and cosmetics launched. One of these was *Milady Face Powder (1)*, "it adheres to the skin giving it that smooth velvety beauty and fresh daintiness which is the greatest charm of youth". The design was still used in 1940.

1

2

The appeal of Colleen Soap (c1910) *(2)* relies on the image of the Irish lass. This toilet soap was made from the ash of plants and pure vegetable soaps.

The German firm of H. Mack made Pasta Mack, a tablet that added fragrance to bath water. This box *(3)* dates from about 1910.

4

3

Skincare products were first made by Potter Drug & Chemical of Boston Massachusetts in 1878. The range of Cuticura products *(4)* with an orange theme come from the early 1900s when

sales were being extended throughout the world. The leaflet wrapped around the bar of soap has directions printed in eighteen different languages, including Arabic, Russian and Chinese.

62

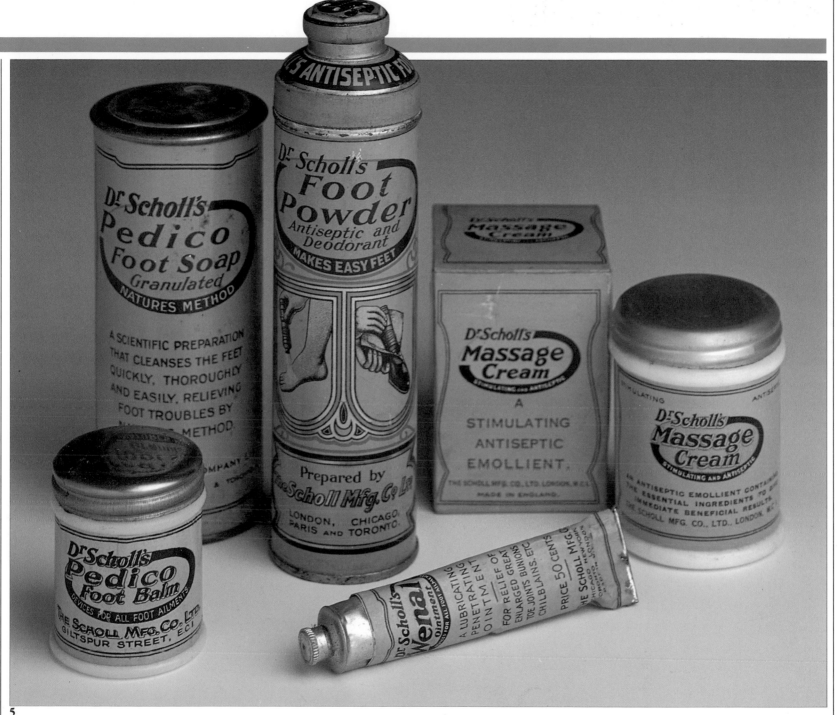

5

William Scholl became aware of people's foot troubles while working at a Chicago shoe shop in 1899. He enrolled at Illinois Medical College and experimented with foot remedies. In 1904 he began to sell an arch support, making them in his own small workshop. The business grew rapidly and by 1910 a factory was being set up in London. The range of foot-care products (5) shown here dates from the period 1910 (Dr Scholl's Foot Powder) to 1920. Each item is decked out in the same livery of a yellow background with a blue oval containing the name of Dr Scholl's product.

TOILETRIES

Aok (*1*) was a German shampoo powder from about 1910. The coloured central label has been stuck on separately. The box depicts the hair styles of twenty-nine different ladies.

1

2

3

The "sweet perfume of Tibet" was captured by the London perfumers J. Grossmith & Son who also incorporated the fragrance into their Tsang-Ihang Sachet (*2*), c 1910. Grossmith specialized in perfumes from exotic places, such as Phul-Nana, "a bouquet of Indian flowers".

The design for the French toilet soap wrapper, Le Suave (*3*) c 1905, was based on the Art Nouveau style.

The toilet soap box lid "Savoly" (*4*), c 1900, was probably drawn by the Czechoslovakian-born artist, Alphonse Mucha, for a firm in Budapest, Hungary. From 1887 Mucha had studied in Paris, coming to prominence in 1895.

4

DENTAL CARE

Kobayashi's Japanese Dentifrice (**5**), c 1915, was manufactured by Tomijiro Kobayashi of Tokyo. The dentifrice was packed in a solid wooden box, for export purposes. The directions were simple, "Apply the powder to the teeth by means of a brush, using moderate friction over the whole surface".

6

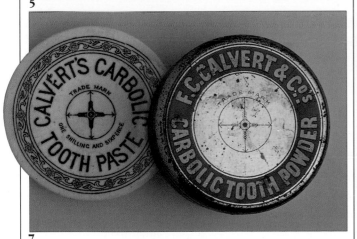

7

Calvert's Carbolic Tooth Powder (**7**) came from Manchester, England. During the 1880s it had been packed in glazed pots, but these were replaced by lighter tins around 1915.

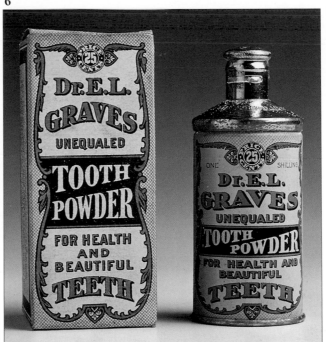

8

Colgate's Ribbon Dental Cream (**6**) was first available in 1908. There were copious illustrated directions on how to squeeze the tube, and how to press the lower sides to return any cream that "oozes", and to fold up the tube as the cream was used. The big advantage over Colgate's previous tubes (used since the 1890s) was that the cream came out like a flat ribbon. It was therefore more economic and it did not roll off the brush.

Dr Graves' Tooth Powder (**8**) used a patent dispensing cap first introduced by Dr Lyon in 1891. Produced in Chicago, the tooth powder was exported all over the world. This pack dates from around 1900.

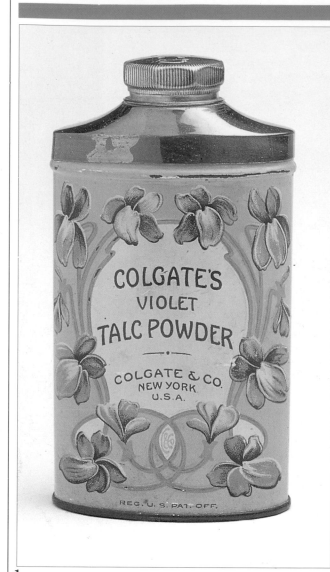

1

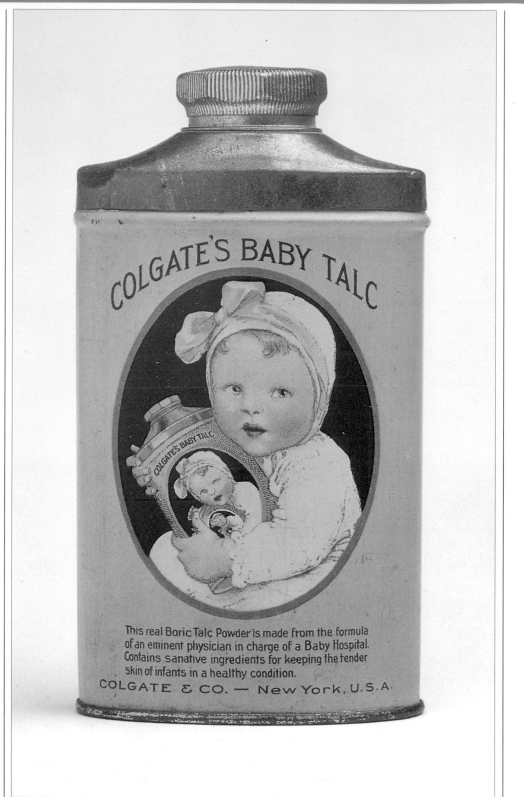

Talcum powder was first packed into tins in the 1890s. These tins rapidly reached a peak in popularity, craftsmanship and visual splendour during the first two decades of the twentieth century.

William Colgate had started in 1802 selling soap and candles in Baltimore. By the 1880s a range of toiletries was being manufactured, which, by the end of the century, included talcum powder. Colgate's Violet Talc Powder (**1**) dates from around 1905 and the design of the tin was influenced by the Art Nouveau movement. Colgate's Baby Talc (**2**), with its diminishing repeated design, was produced from around 1910.

2

3

One early manufacturer of talcum powder was Gerhard Mennen of New Jersey. In 1889 he brought out Mennen's Borated Talcum Toilet Powder in a cardboard drum. Unfortunately this leaked, so the Somers Brothers were commissioned to develop a tin for it. The next year a cylindrical tin was introduced with a sprinkler top that could be revolved to open or close.

The flattened oval shape (**3**) replaced the drums in 1909 (the baby design had been devised earlier in 1889). The baby appealed to mothers (the model for the infant was the child of a local supply salesman) while the portrait of Mennen added his personal endorsement to the product.

4

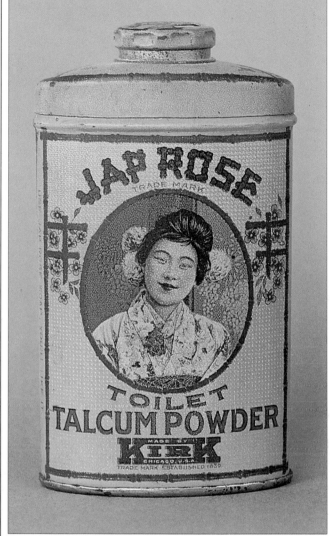

5

Competition amongst talcum powder manufacturers was intense during the first quarter of the twentieth century. One talcum powder was much the same as the next, so firms concentrated on the perfume and the mystique of the brand name along with the tin's decoration. Such was the case with Violet Sec Talcum manufactured by Richard Hudnut (**4**) and Jap Rose Toilet Talcum powder manufactured by Kirk of Chicago (**5**), American products of about 1905.

SPICES AND FLAVOURINGS

In 1868 Joseph Watkins started to bottle various medicinal products, and sell them from a horse and buggy around the locality of Plainview, Ohio. The business grew and in 1885 he moved to Winona, Minnesota.

By the turn of the century, the range of products was extending to food products like cocoa, pepper and mustard. The pack here (**1**) dates from around 1910.

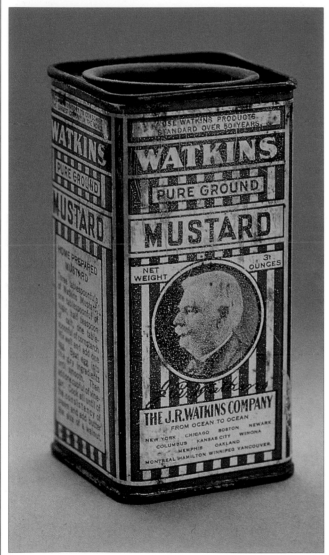

1

3

Alvin Woolson set up the Woolson Spice Company at Toledo, Ohio, in 1882. He was a pioneer in selling roasted coffee beans individually wrapped for the customer; at that time coffee beans were sold from open bins. The Golden Sun Spices design (**2**) was registered in 1912.

It was in 1847 that Baron Justus von Liebig, a German chemist, developed a concentrated extract of meat. During the 1860s there was a vast surplus of meat in South America which he was able to use. Leibig's Extract arrived in Britain in 1865. In 1910 an Oxo Cube (**3**) was developed. Each cube was wrapped in a card box, being sold for one penny. The Oxo tins had a background pattern of red cubes, a design which remained the same for 35 years.

Robert French started a spice business in 1880. Five years later he moved from New York to Rochester where he started the manufacture of mustard. The design (c1905) for French's Pure Grated Nutmeg (5) is held together by three interlinking bands that have an Art Nouveau flow.

Frank's Red Pepper tin (6) (c1910) relies on the striking colours of the Italian flag to gain attention. "Insist on Frank's" is boldly inscribed on the lid.

5   6

7

It was Frederick Garton of Nottingham who first made HP Sauce (4) in the 1870s. His secret recipe was purchased by the Midland Vinegar Company and in 1903 HP Sauce was relaunched. It has never been established whether HP stood for Houses of Parliament, but since 1903 the label has had the famous landscape bestowed upon it.

At this time, Slade's Ginger (7) was another product to accentuate its purity.

PROVISIONS

This is an early example, c 1900, of a product character, drawn by the French artist, Farcy, for Père Lustucru egg noodles (*1*).

Around 1910 Punch Custard Powder (*2*) was produced by John Connell & Co of Melbourne and Sydney, Australia, who also manufactured jams.

1

2

3

This Normandy Camembert label (*3*) dates from the early years of this century. The earthy colours of the design owe much to the influence of the Art Nouveau movement.

Before the peanut butter pail became popular in the USA during the 1920s, the French had sold jam in novelty tin pails, such as this one, c 1910, from Marseille (*4*).

4

SOFT DRINKS

In 1886 Dr John Pemberton, a pharmacist in Atlanta, Georgia, first sold Coca-Cola at five cents a glass. Three years later all rights to the product had been bought by Asa Chandler, owner of a pharmaceutical company. Initially mainly sold at soda fountains, Coca-Cola was first bottled in 1894, and many firms later bottled it under authority. The classic Coca-Cola bottle (6), created by Alex Samuelson of the Root Glass Co, arrived in 1915.

Another pharmacist, Caleb Bradham from North Carolina, invented Pepsi-Cola in 1898. The bottle shown here (8) dates from 1950 when the logo was put in an oval shape, but the scripted logo had been used from the outset and was modified in 1906, to the form shown here.

6

Canada Dry Pale Ginger Ale was the invention of John McLaughlin, a chemist from Toronto who, in 1904, greatly improved upon the colour and flavour of existing ginger beers. Canada Dry (5) was launched in the USA in 1923.

In 1873 T. H. W. Idris founded a soft drinks company in England. This heavily embossed miniature bottle of Idris Fruit Syrup (7) would have been available between 1910 and 1925.

5

7

8

BEVERAGES

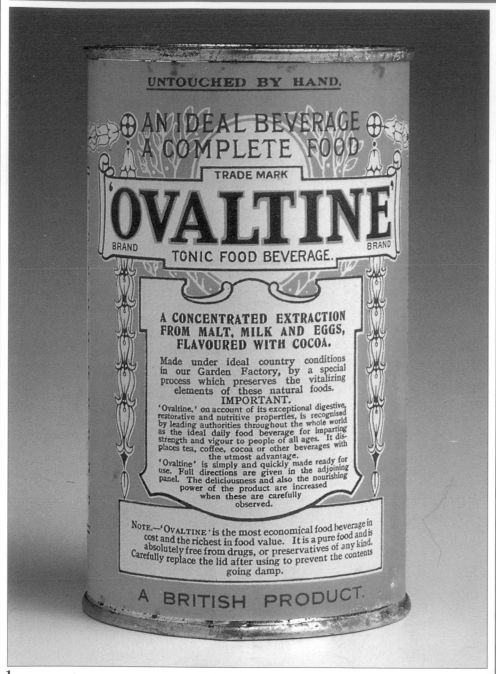

1

This American tin of Chase & Sanborn Formosa Ooloong tea (**2**) *dates from c 1910. It has an uncomplicated design.*

2

*The product first marketed in 1904 as Ovomaltine was created by Dr George Wander, a Swiss chemist. Later, as Ovaltine (**1**), it became popular throughout Europe. The British*

company, A. Wander Ltd, was formed in 1909, and a factory was built at Kings Langley four years later. The Ovaltiney Club promoted through Radio Luxembourg arrived in 1935.

*Joel Cheek created a special blend of coffee in 1892 and named it Maxwell House (**3**) after the hotel in Nashville where it was first tested. This design probably dates from around 1910.*

3

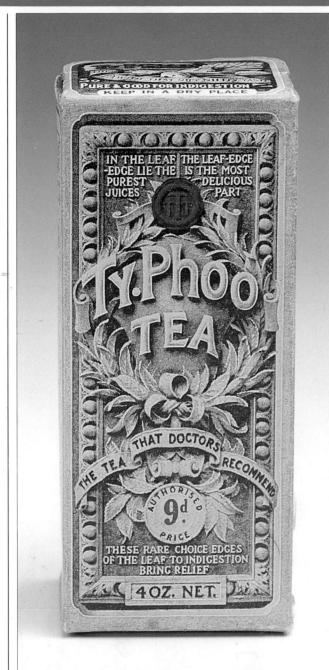

Peacock Brand Tea (**5**) was made in China using the choicest Keemun tea, but was produced and packed by the China Tea Company of Hong Kong. Most of the inscriptions are written in Chinese. The design dates from around 1915.

4

Originally called Sumner's Ty-phoo Tea (**4**) after its creator, the foliage-adorned grey carton was launched in Birmingham, England, in 1905. Billed as bringing relief for indigestion, it was originally purported to be the "tea that doctors recommend". However, medicinal claims were banned in the 1940s. By the 1970s much of the foliage design had disappeared to reveal the now familiar red pack. An illustration of teapot and cups has recently been added.

5

Menier's Powdered Chocolate was first produced in Paris in 1816; its popularity became world-wide, with a London factory being opened in 1863. The image of a girl writing on a wall, drawn by the French artist, Firmin Bouisset, in 1893, appeared on many Menier packs, as also on this American tin (*1*) of around 1910.

This pack would have contained twelve tins of Plasmon Cocoa (*2*), "one cup contains more nourishment than 10 cups of any other cocoa". The design, dating from around 1905, has a simple Art Nouveau theme.

The spicier cocoa produced by the Dutch firm, Van Houten, was gaining popularity over Cadbury's Cocoa Essence, so in 1906 Cadbury's launched Bournville Cocoa (*3*). This brand, with the favoured spicier taste, soon outsold the old Cadbury Essence. The orange background to Bournville Cocoa has remained ever since.

3

2

The Chocosubit (*4*) tin using flower motifs in both foreground and background also dates from around 1905. The manufacturer, Julien Damoy, claimed that the chocolate drink could be prepared in 10 seconds.

4

HOUSEHOLD CLEANERS

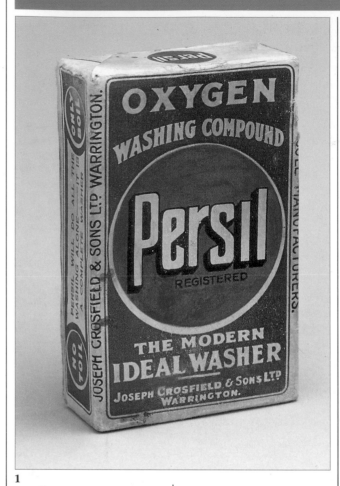

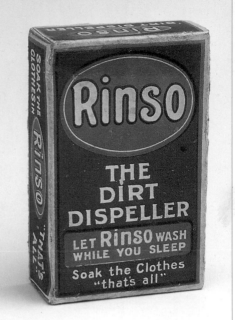

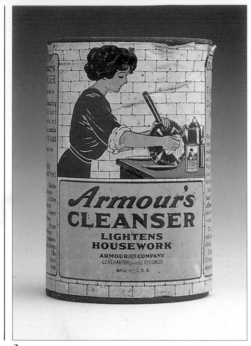

1

Manufactured from 1907 by the German chemical firm of Henkel, Persil (1) was first produced in England in 1909 by Joseph Crosfield of Warrington. One advantage of Persil was that it contained a strong bleaching agent, capable of removing most stains.
The public was encouraged to believe that Persil did the washing on its own – "it is a complete washer: no rubbing or scrubbing" and dispensed with the effort of the "dolly" and washboard.

Another meat-packing company was that of Cudahy, Nebraska, founded in 1887. They, too, had launched a scouring powder in 1905, called Old Dutch Cleanser (4). This symbol of a Dutch girl brandishing a stick (and chasing a goose) was copied from a decoration found painted on a picture frame.

4

The firm of R. S. Hudson of Liverpool was first established in the 1830s. Rinso (2) was launched in 1910 to compete with Crosfield's Persil. Although Lever had purchased Hudson's in 1908, the firm kept its separate identity.
Rinso was advertised as giving perfect results, "Let Rinso wash while you sleep". Curiously, the packet also stated that Rinso made a perfect hair shampoo.

The origins of Armour & Co go back to 1863 when Philip Armour and John Plankinton started a provisions business in Wisconsin. They moved to Chicago four years later and went into meat processing, and then into meat canning by 1879. Branching out into a new product area, they launched Armour's Cleanser (3) around 1905.

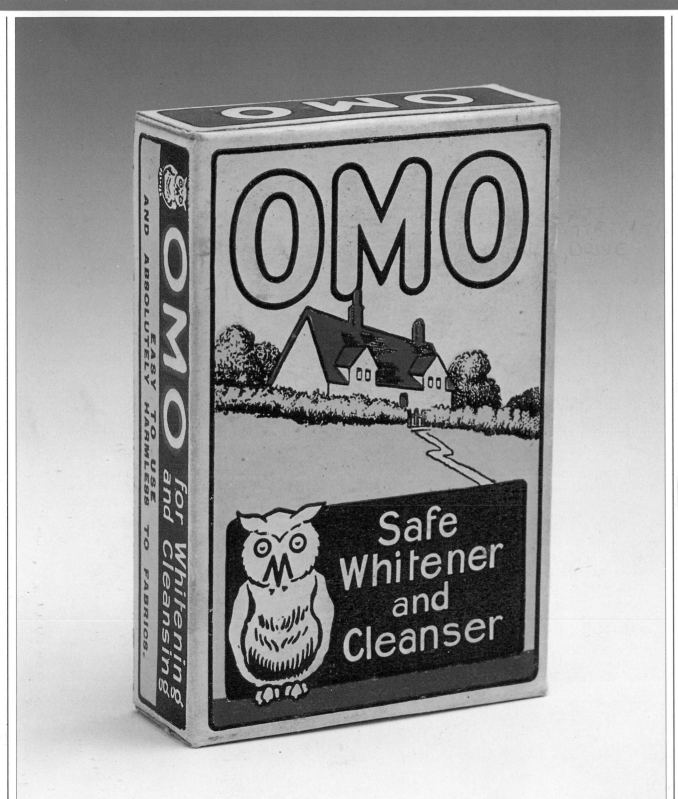

Omo (**5**) was another cold-water washing powder like Persil and Rinso. "Science has now discovered a skilful preparation of oxygen which offers in Omo the healthy, whitening, cleansing and purifying properties of the fresh country air in a concentrated and convenient form." The pack reflected this and the letters Omo were cleverly placed in the wise owl's face. Omo was launched by R. S. Hudson in 1908 but the brand did not succeed until it was relaunched in 1954.

The Australian cleaning powder, Sanoper (**6**) (c 1910), was also distributed in Britain. It did not have the usual sprinkler top – instead the lid turned until a slot appeared.

5

6

BOOT POLISH

A liquid boot blacking had been the main shoe-cleaning product in the nineteenth century. But in the early twentieth century solid paste was introduced and became the preferred polish thereafter.

Cherry Blossom Boot Polish (**1**) goes back to 1903, being an invitation of the Mason brothers of London. Crème Eclipse (**3**) by Cirages Français appeared around 1900. The eclipse motif was used on all their advertisements.

1

2

3

4

Marketed first in Australia, from 1906, Kiwi Boot Polish (**2**) was named by its creator William Ramsey in honor of his wife, a New Zealander. In 1912 Kiwi was sold in England, particularly to the British Army. The demand during World War I helped the brand to take off, and eventually sell around the world.

The design for Radium Polish (**4**) of Manchester, England, probably dates from around 1915.

MEDICINAL

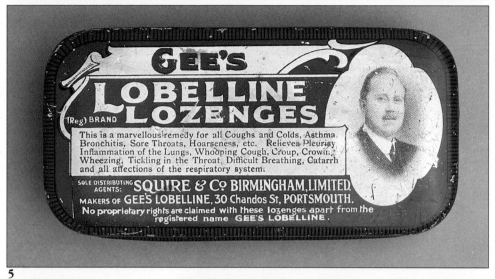

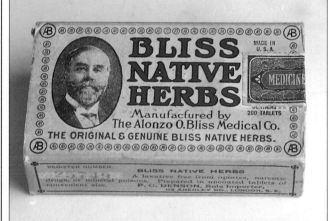

Gee's Lobelline Lozenges (5) from Portsmouth, England, claimed to be a marvellous remedy for all coughs, colds, asthma, whooping cough, bronchitis, hoarseness, pleurisy, wheezing and catarrh.

Bliss Native Herbs (6) – a laxative manufactured in Washington, USA – were exported to Britain during World War I. One shilling was donated to the War Funds for every box sold.

6

7

The American design for Haywood's Powder (7), manufactured by Pfeiffer Chemical Co, probably dates from around 1910. The borders are still intricate and have been influenced to some extent by the Art Nouveau movement.

8

A French laxative, Purganol (8), produced by Daguin, was packaged in a cylinder which was placed in a three-cornered box. An advantage of this pack was that the label could be read at a glance. The name and address has been repeated to form a background to this design, dating from around 1910.

9

Mentholatum (9), the cure for colds, cuts, insect bites, headaches and aching feet, was sold in the USA, Britain and other countries around the world. The small tin here was a travelling pack (c 1915) which could be refilled from the larger container.

CIGARS

FLOR DEL REY

Superior SEGARS
SPECIALLY MANUFACTURED
for GENTLEMEN OF
REFINED TASTE.

Superior SEGARS
SPECIALLY MANUFACTURED
FOR GENTLEMEN
of refined taste.

1

*Cigar labels have traditionally been of the highest printing quality, embossed with gold highlights. One such label was that for Flor Del Rey (**1**), having none other than King Edward VII as its "sponsor" in the centre. It was unusual to have any writing on the label except for the brand name, but here the manufacturer could not resist stating, "superior segars specially manufactured for gentlemen of refined taste"*

*Not all cigars were large, although the best labels usually went with the bigger cigars. This tin of Red Dot Junior Cigars (**2**) was good value and the price was made prominent. They were made by Barnes-Smith Co of New York State around 1910. A lady's face beckons from the centre of the red dot from which the sun-ray motif emanates.*

FOR

10   50¢

RED · DOT

JUNIOR CIGAR
BARNES-SMITH CO., MAKERS.
BINGHAMTON, N.Y.

TRULY DIFFERENT

2

John Freeman started to make cigars in 1839, initially in London and later in Cardiff, Wales, where they moved in 1895. Around the turn of the century cigar smoking had become more popular in Britain, in particular the cheaper smaller cigars that were starting to be made by machine. In 1912 J. R. Freeman & Son launched Manikin Cigars (**3**), a small cigar with a mild Havana flavour packed in a pull-and-slide carton. The brand name had been inspired by a music hall act, The Three Manikins, that Peter Freeman had seen.

3

De Fox (**4**) was a Belgian cigar brand; this label dating from around 1900 was influenced by the Art Nouveau movement.

The design for Miss Butte (**5**) was registered for copyright by the American Lithographic Company in 1902. A continuous stream of new stock designs was created by their studios. The cigar manufacturer then chose which one he liked, buying them at the rates quoted here – $24 per thousand or, in Britain, £3 12s per thousand.

4

5

EXOTIC CIGARETTES

The firm of B. Morris & Sons of London specialized in cigarette tobaccos from all over the world – Havana, Russian, Latakia, Mexican, Virginian, Rhodesian and particularly Turkish and Egyptian (*2*). Exotic-looking packs, designed about 1905, were used for the cigarettes.

1

2

R. J. Reynolds had been in the tobacco business since 1875 in North Carolina. In 1890 the five leading tobacco manufacturers in the USA formed the American Tobacco Company, but in 1911 the conglomerate was split up by law. In 1913 Camel Cigarettes (*1*) appeared, following a "teaser" campaign which announced "The Camels are Coming". The Camel image was actually based on a dromedary called Old Joe, of Barnum and Bailey's Circus. (The Circus happened to be in town when the designers were looking for a model.) Camel cigarettes were an immediate success and the design continues in the same fomat.

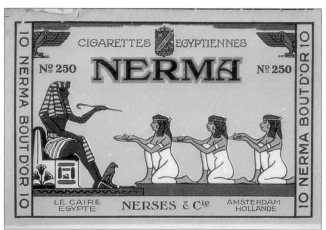

3

Sambul Cigarettes (*3*) were issued by the Orient Cigarette Company of Cairo, Egypt, and date from about 1915.

4

Another cigarette from the same period with an Egyptian theme to the pack design was Nerma (*4*), manufactured by Nerses & Co of Amsterdam, Holland.

Between 1900 and 1930 Turkish and Egyptian cigarettes were in vogue. Although little tobacco was grown in Egypt (Turkey, Greece and the Balkans grew most of the exotic tobaccos), ancient Egyptian culture had a certain glamour and the opportunity to exploit it for marketing purposes was too good to miss. Sakkara Cigarettes (**5**) had an Egyptian blend tobacco, as did Padapoulos Frères (**6**), a firm founded in 1881. Both designs date from around 1910.

5

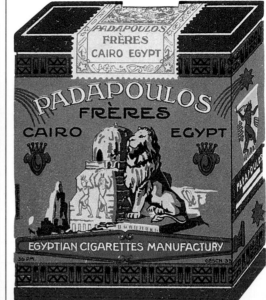

6

Muratti's Neb-Ka Cigarette tin (**7**), c 1900, shows a scene from Constantinople. B. Muratti & Sons were a respected London company, but, as with many other reliable firms, they suffered from the activities of those producing similar packs with inferior quality products. Muratti added a warning on the back of their tins, "We therefore caution buyers that none are genuine unless our name, Muratti Au Bon Fumeur, appears on the label outside, and in circulars inside each box".

7

TOBACCO PRODUCTS

In 1917 the American Tobacco Company, founded by James Buchanan Duke, launched Lucky Strike Cigarettes (*1*). They took the red bull's eye motif from Lucky Strike Tobacco – a brand established since 1871. The pack remained basically the same until 1942 when the green background was replaced with white.

In the early 1900s Americans realized that tobacco tins would be better if they were both flat for the pocket, and had a small lid at the top. This enabled the pipe to be filled with the least risk of spilling the tobacco. An example of this new design was the tin for Velvet Tobacco (*2*) manufactured by Liggett & Myers of St Louis, later known for their Chesterfield Cigarette, launched in 1912.

1

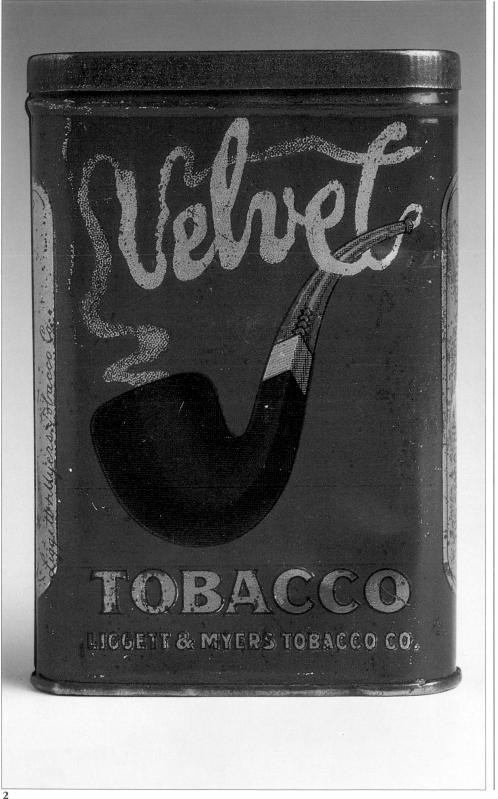

2

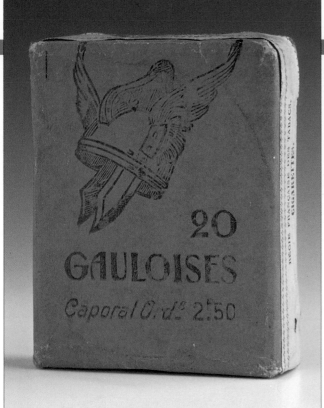

3

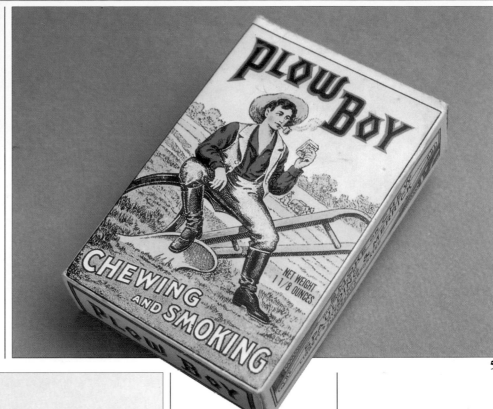

5

4

From Napoleon's time the French Government has controlled the production and sale of cigarettes in France. Just before the outbreak of World War I, Gauloises (*3*) were launched in a soft pack. The Gallic war helmet symbol and blue wrapper is still used.

Founded in 1788, the London tobacco firm of Carreras had registered their black cat trade mark in 1886. This pack of Black Cat Cigarettes (*4*) had a green band, denoting that it was the extra large variety, made around 1910. The pack also enclosed coupons. Gift schemes were dropped in Britain in 1934, but returned in the 1950s.

The chewing and smoking tobacco brand of Plow Boy (*5*) was manufactured by the American firm of Spaulding & Merrick. Probably a late-nineteenth-century brand, the graphics have been tidied up for this 1910s version.

*Optrex Eye Lotion was the creation of a
Frenchman, M. Rosengart, in 1928.
Optrex was introduced to Britain in
1931. The eye bath, which sits over the top
of the bottle, was originally made from
glass; but since the 1960s a plastic one has
been used.*

CHAPTER THREE
DECO STYLES

1920-1939

INTRODUCTION

The changes brought about by World War I did much to jolt the world into a new era. The greater participation of women in the commercial world, the gradual passing of the domestic servant and the increase of leisure time, were all social changes that helped to generate a new attitude toward packaging design.

The package designs of many branded goods had remained substantially unchanged for forty or more years. The public had grown to trust the contents; the manufacturers for their part had been happy to let the continuity speak for the fact that the contents remained reassuringly the same and of consistent quality. However, when new brands did arrive, they tended to take on the modern design idiom. The appeal of a product in, for example, an Art Nouveau livery would give it a selling edge over a similar product packaged in a more traditional style. The implication that the contents were fresher or better or more up-to-date could not be ignored.

As a consequence much rationalizing of pack design took place during the 1920s and 1930s. Established brands tended to have their pack design modified slowly over a period of time and often the changes were imperceptible. Even so, many packs escaped altogether from the designer's brush of modernism (it was often the founder of the firm who prevented change). A marketing psychologist said in 1936 that "most manufacturers have an exaggerated idea of the amount of goodwill attached to a package design that has been associated with their firm for a long period of years". The products that survived in Victorian dress were usually those associated with a particularly characteristic taste, such as a sauce, or the perceived flavour of a traditional tobacco. (For example,

In 1924 the American tin manufacturer Tindeco commissioned the illustrator Harrison Cady (known for his popular Peter Rabbit cartoons in the New York Herald Tribune*) to create a range of Peter Rabbit tinware. Amongst these was a peanut butter pail (1) and a talcum powder tin.*

The Holcomb & Hoke Butter-Kist Pop Corn pack (2) was printed in a single colour making it economic to produce. Different recipes were printed on four of the carton flaps. Butter-Kist was made in Indianapolis and then distributed to retailers who popped the corn in the Butter-Kist machine.

2

the designs of Woodbine and Player's Medium cigarettes remained unaltered until 1960.)

Changing a packaging material had (and still has) other implications apart from jeopardizing the continuity of consumer acceptance. There was much debate over whether or not to change from one packaging material to another – for example, from pottery to glass, from glass to Bakelite, or from tin to aluminium. Any change of this nature could involve expensive new packaging machinery. Furthermore, there were purchasing habits to be considered. Financial depression, the development of new types of outlet, changing social habits and different uses to which the product could be put were all important in affecting how the package might be reformulated. For instance, there was a trend for smaller pack sizes, partly due to smaller family units (without servants), partly from financial restraints of the Depression which meant that less could be afforded at one time, and partly from the requirements of the new fixed-price stores like Woolworths (where nothing was priced over 6d. or 10 cents).

Much of the need for cleaner, clearer design was

3

inspired by the new wave of styles that broke with the past. The Art Nouveau movement had begun at the end of the previous century. By the 1920s a distinctly different style had evolved, now known as Art Deco: strong geometric shapes coupled with vivid colours were its hallmarks. As this exciting new style swept the design world, the result was that packaging graphics became bolder, dropping the fussy and ornate style which had marked earlier pack design.

The first types of products whose packaging style was affected by the Art Deco movement were toiletries and cosmetics (just as they had been with Art Nouveau). Some firms had already embarked on clear labels, such as Elizabeth Arden with its Venetian range in 1915 and Chanel with its No.5 range, launched in 1921. Cosmetic packaging had always tended to move with the latest design fashions, partly because it is closely linked to the world of *haute couture*, but also because so many competing brands continued to be launched every year. It is interesting to note that when a Frenchman wrote an article in 1928 on contemporary French packaging, he was unable to make a general survey of artistic packaging of

goods in France: "I must confess that whereas articles of luxury are presented in a really artistic get-up, the articles of daily life usually appear on the market in very mediocre wrappings. All provisions, for example, are sold in wrappings as like as peas. The wrappings created for articles of quality, on the contrary, are extraordinarily interesting, numerous and often extremely original." His comments on the reasons for this were that cosmetic packings were "in a certain sense a part of the article itself..... judged according to the outward impression".

Design apart, there were major advances in the field of packaging technology during this period, particularly in plastics. In fact, the search for synthetic substances had been developing throughout the nineteenth century. In 1869 the Hyatt brothers of New York had discovered a non-brittle material made from camphor and cellulose nitrate, which they named Celluloid. In 1907, the Belgian-American chemist, Leo Baekeland, discovered the first truly synthetic plastic which he named Bakelite. It was during the 1920s and 1930s that this material was most used, frequently in brown but also in dark shades of red, blue, green and

INTRODUCTION

1

The use of new packaging materials always intrigued the public. The mottled effect of Bakelite was no exception, although it was never used extensively for household products. The example here, Tooth Soap (1) of about 1930, was produced by the British firm Enolin Limited, founded in 1926. The block of dentifrice inside was wrapped in Cellophane.

American slicing machine was imported to Britain in 1929.

Further developments were also being made with aluminium containers. The advantage of aluminium lay in its pliability, its attractive finish, and its light weight – a third of the weight of tin. It was, unfortunately, three times as expensive. Probably the brand which best utilized the aluminium "tin" was Gibbs Dentifrice, which relied entirely on the brightness of the aluminium to give it sales appeal. (In 1947 the container was updated, the customer now being given a choice from three different coloured aluminium tins.) Although solid dentifrice and tooth powders had been popular for some time, toothpaste in a tube had started to dominate the market by the 1930s. Many other products were increasingly using the

black. Many further developments and refinements followed, allowing moulded plastic containers to be used more widely, particularly for toiletries such as shaving stick cases and cosmetic containers. However, the high cost of using plastics for packaging prevented their wider acceptance at this stage. Nevertheless, during the 1930s, plastic table- and picnic ware, light fittings, telephones and even coffee grinders could be purchased in this modern material.

Another substance that became widely used in the 1930s was Cellophane, a thin transparent film that could be wrapped around any pack to seal it – making it more hygienic and keeping the contents fresher. A Swiss inventor, Jacques Brandenberger, had discovered the formula for this material as far back as 1869 but it was not until 1913 that it was first manufactured, under the name of Cellophane. By the 1920s these transparent sheets were being used in the confectionery industry to wrap boiled sweets individually and, later, to wrap a variety of cartons and boxes such as cigarette packets. (The first brand of cigarettes to do so in Britain was Craven "A" in 1932.)

Although Cellophane was able to replace a whole range of greaseproof and waxed papers, it did not take over from the traditional waxed bread wrappers. These had been used in various ways prior to the first successful bread wrapping machine introduced in America in 1913. In the 1920s the use of bread wrappers increased particularly when a bread slicing machine was developed – it was, of course, a necessity for selling the sliced loaf. The

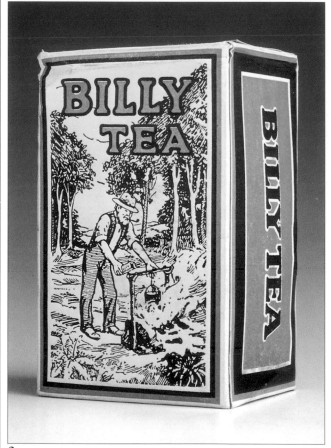

2

Billy Tea (2) was introduced to Australians in 1888 by James Inglis & Co of Sydney. On one side of the pack the pioneer tends his billycan of tea, on the other a

kangaroo carries a bedroll over his shoulder and holds a billycan. This modified pack dates from the 1930s.

3

4

The invention of the beer can in 1934 revolutionized drinking habits. Two types were available: flat tops that needed a special can opener and cone tops that required only a standard bottle opener.

Budweiser Lager Beer was brewed by Anheuser-Busch of St. Louis, a brewery which started in 1852. Budweiser itself was launched in 1876. The can here (**3**) dates from the late 1930s; it was Keglined and diagrams on the side show how to open the can with the Canco "quick-and-easy" opener.

The Duquesne Brewing Company of Pittsburg produced their Can-o-Beer (**4**) in 1935, with the following explanation on the back: "This cap-sealed can brings Duquesne Beer to you with its outstandingly fine flavor unimpaired. Light cannot affect it. It visits no home but yours. It cools quickly and takes up little space. Open it like a bottle – drink from it if you wish. When emptied, crush it and throw it away. No empties to return. No 'extras' for deposits."

"collapsible" tube – shaving cream, face and hand cream, glue, shoe polish, and a variety of foodstuffs such as meat paste, cream cheese and condensed milk. However, these edibles were found to be acceptable only on the Continent and in the United States; it would appear that British sensitivity for the moment could not cope with fish paste squeezed from a tube.

Waxed cardboard cartons had been used as containers for cream since the 1910s, but by the 1930s their use had widened to include honey, glacé cherries and ice cream. A further development was the use of waxed-paper cartons for milk, referred to at the time as "paper bottles".

Glass bottles also came under attack in the "take-home" beer market. Canning companies had the novel idea of packaging beer in what they claimed was a superior container to a bottle – a can. Cans were easier to stack, took up less space and could be filled more quickly than a bottle. Although cans could be used only once, returnable bottles involved the troublesome administration of deposits and recycling. However, the most important point was how good the canned beer tasted. In 1934 the

American Can Company patented their beer can using the trade mark, Keglined. In the following year a number of American brewers began selling canned beer; it was an immediate success with the drinking public, still jubilant from the overthrow of the prohibition laws in 1933.

Inevitably there were some problems with cans. Brewers had to install expensive machinery for the filling of these new cans which had flat tops; they also required a special opener. In the same year Continental Can devised a can, the cone-top, that mimicked a glass bottle and it could use existing filling machines. Sealed with a Crown top, it could be opened with a standard bottle opener. It was this latter type of can which was used first in the European markets. The cone-top can did have one unfortunate point, namely its similarity to a can of metal polish, and, as might be expected, met with consumer resistance.

In the story of packaging, there is often a simple solution to a niggling problem. One such problem was how to lever off the lid from a flat tin of polish. The solution was the simple addition of a "winged" lever which, when twisted, effortlessly raised the lid.

ANIMAL IMAGES

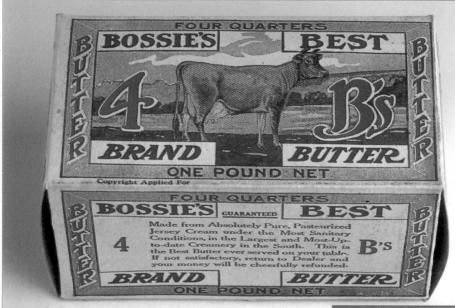

1

Bossie's Best Brand Butter (**1**) came in a waxed card carton, issued by the Aberdeen Creamery Company, USA, around 1930. The Jersey cow indicated it was a quality product, but this was also backed up by the statement that the dairy had "the most sanitary conditions, in the largest and most-up-to-date creamery in the South". A further claim was made: that each pound of butter was worth, as an energy producing food nearly four times as much as a dozen eggs and eleven times as much as a pound of fish.

The humorous image adopted by Laughing Cow Cheese (**3**) and created by Benjamin Rabier, around 1930, helped to give the cheese its great popularity.

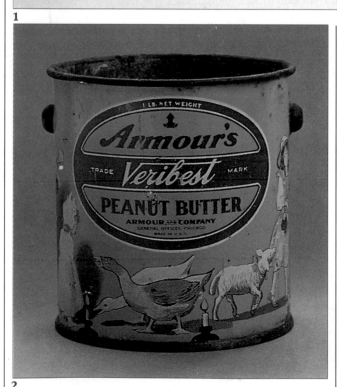

2

Armour's Peanut Butter tin pail (**2**) of the 1920s bore a design incorporating farmyard animals – an especially appealing image for the children who consumed it in large quantities.

3

5

The traditional box for packing dates has always been similar to this one for La Girafe (**4**), c 1930. The dates from Tunis were packed in neat rows with the stem down the centre. Today, plastic boxes have replaced the original wood and card ones, and imitation plastic stems now divide the rows of dates.

Libby's Evaporated Milk can (**5**) of the late 1930s emphasized that the Vitamin D content had been increased. The label design for this can had been registered in 1925.

4

PROVISIONS

A Canadian, James Kraft, came to the USA in 1903 to start a wholesale business. Later he began to produce processed cheese. Business boomed when he was contracted, in 1917, to send cheese to the troops overseas. Foil wrappers for 5lb cheeses were used from 1921, replacing tins. This Kraft Velveeta box (1) is British, designed in 1935.

After some experimenting, William Wright, a food salesman in Chicago, developed a better form of baking powder. He started to sell it in 1890 under the name Calumet (the word used by the French for the Indian peace pipe offered to the explorer Père Marquette in 1675). This stylized design for Calumet (2) dates from the 1930s.

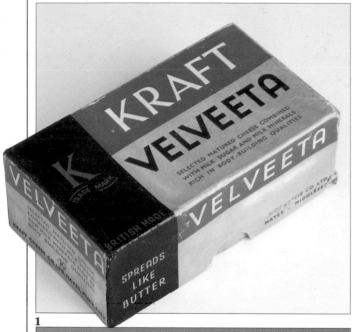

1

2

3

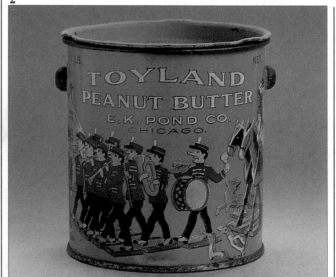

Tin pails filled with peanut butter were popular with children in the 1920s. Not only was peanut butter a treat, but the colourful pail made a useful plaything afterwards.

When Jackie Coogan became popular following his success in 'The Kid' (1921), his picture appeared on peanut butter pails produced by the Kelly Company of Cleveland, Ohio (3). Toyland Peanut Butter (4) was produced by E. K. Pond & Co of New York.

4

By the 1930s the largest group of retailers in Britain were those of the retail co-operative societies. The movement started in 1844 at Rochdale, Yorkshire, and was supplied by the Co-operative Wholesale Society (CWS). Canned foods had at last become popular in Britain; modern graphics added to their appeal, as with CWS Peas and Butter Beans (**5**).

In 1869 Joseph Campbell had started to can foods in New Jersey. Canned soups arrived in 1898 sporting the gold medallion on a distinctive red and white label (allegedly inspired by the colours of the Cornell College football team). The can here (**7**) dates from the 1920s.

5

Alfred Bird, an experimental chemist in Birmingham, England, first produced an egg-free custard powder in 1837. Later, during the early 1870s, he developed and sold a Blanc-Mange Powder, which at that time was a novel cold dessert. Originally sold in a box without any illustration or colour, with just the name and directions on the front, it was not until around 1910 that the packs were made more colourful. The pack here (**6**) dates from about 1930.

6

7

SNACKS

Potato crisps have been a favourite with Americans since the 1850s but it was not until 1920 that Frank Smith found just the right crispness and introduced Smith's Potato Crisps (1) to Londoners. The packets sold for 2d each, and a twist of salt was included for those who wished to add it. Larger quantities of crisps were sold in tins, an ideal way of keeping the crisps fresh for a picnic or party.

For many years the National Biscuit Company (NBC) had been looking for a buttery cracker to compete with others already being sold. In 1934 a new cracker was introduced, different from its competitors in that the flavour was enhanced by a thin coat of coconut oil and a sprinkling of salt. For this prestige item the name Ritz (2) was chosen. Within three years of the launch, 29 million crackers were being baked a day.

1

2

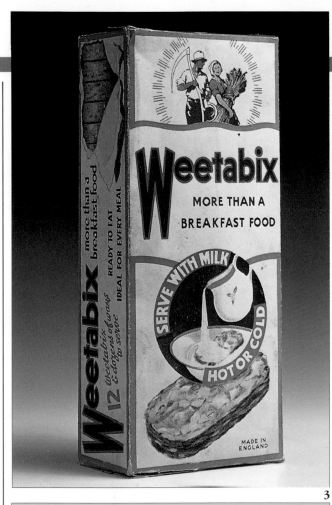

3

4

6

In 1932 four South African Seventh Day Adventists set up the British and African Cereal Company to sell their product, Weetabix (*3*), in Britain. This new wheat cereal was not just recommended for breakfasts, but also for light lunches, "cut lengthways with a knife ... eat them with plenty of fresh butter, jam, cream, honey, or cheese."

Shredded Wheat (*5*) was sold in Britain from 1908, but it was not until 1925 that a factory was built at Welwyn Garden City. Again, the packet depicted the modern factory, described as "a palace of crystal, its great walls of glass held together by slender white-tiled columns". The name Welgar was added in 1941, following the passing in 1938 of the Trade Marks Act.

NBC's Social Tea Biscuit was one of the first sweet biscuit lines to be wrapped in its own carton, in 1899. The design of this pack (*1*), dating from about 1930, has been modified to incorporate a picture of the biscuit on the side.

Weston Clix Cocktail Wafers (*6*) were made by the Weston Biscuit Company at Slough, London, and in Edinburgh, Toronto and New Jersey. This British tin dates from the late 1930s.

5

les
belles
de
Montmorency

Rozan

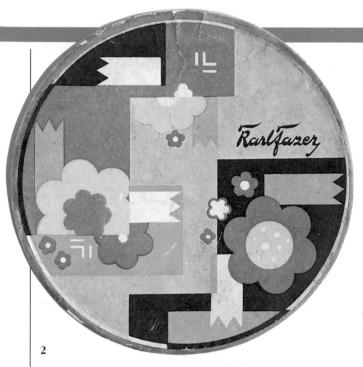

2

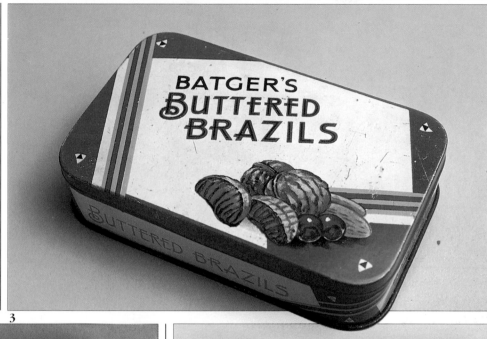

3

4

5

The French confectioners Rozan (**1**) manufactured their chocolates in the Pyrenees. This box of chocolate cherries called "Les Belles de Montmorency" (after a noble French family) was produced in the 1920s and has a remarkable label printed in such a way as to appear like the original painting.

This Karl Fazer (**2**) chocolate box from Helsinki, Finland, decked out in Art Deco style, dates from about 1935. Henri Nestlé founded his milk product company in 1866 at Vevey, Switzerland. A neighbour, Daniel Peter, was a chocolate manufacturer. In 1875 they produced the first milk chocolate. By 1905 Nestlé's chocolate had begun to be distributed widely in Britain. Nestlé's Fruit Creams (**4**) arrived in the 1920s.

This tin of Batger's Buttered Brazils (**3**) has a geometric Deco style, dating from around 1925.

The imitation cigarette has long been a popular sweet amongst children. Candy Smokers (**5**) were made by Winnie Connie Specialties of Wisconsin, in the late 1930s.

CANDY AND CHOCOLATE

1

These three Chocolate Sandwich (**1**) bars are from a series of six produced by Cadbury's of Birmingham in the late 1920s. The cartons were designed by Austin Cooper, best known for his posters, but also celebrated for his use of bold lettering.

In the 1890s Milton Hershey was making caramels, but he switched in around 1905 to making milk chocolate bars at a large factory in Pennsylvania. They sold at a modest five cents a bar and, unlike similar bars, were individually wrapped. This bar (**2**) dates from 1935.

2

3

These two French chocolate bar wrappers (**3**) in Art Deco style date from around 1930.

Black Magic (**4**) was launched in 1933 by Rowntree's of York after extensive market research had gone into finding which chocolate centres were most suited to British tastes.

4

In 1862 Henry Rowntree took over a small cocoa manufacturing company in York. Crystallized gums were first produced in 1881 and clear gums in the 1890s. At this time the company had also started to produce boxed chocolates. With the increasing popularity of chocolate bars in Britain, Rowntree's launched Chocolate Crisp in 1935. It was renamed as Kit Kat two years later (**6**).

6

5

The 1930s saw great popularity for candy bars (**5**). In America the firm of Luden's sold 5th Avenue; Baby Ruth was produced by the Curtis Candy Co of Chicago from 1920; and Oh Henry! was the creation of George Williamson, also of Chicago. All these bars sold at five cents each, as did those from Frank Mars. He began making candies in his kitchen in 1911. In 1920 he set up a factory at Minneapolis. Milky Way was launched in 1923 and the Mars Bar in 1930. Maltesers (**7**) arrived in 1936. It was Forrest Mars, his son, who set up a factory in 1932 to produce Mars Bars in the UK.

7

TOBACCO PRODUCTS

Tobacco tins to fit the pocket were favoured by Americans, particularly between 1905 and 1935. Hi-Plane Tobacco (*1*) was manufactured by Larus & Brother of Richmond, Virginia, and dates from about 1930.

This box of Polish Dames Cigarettes (*2*) would have been designed in the late 1920s, and aimed at women smokers, a small but growing part of the cigarette market in those days.

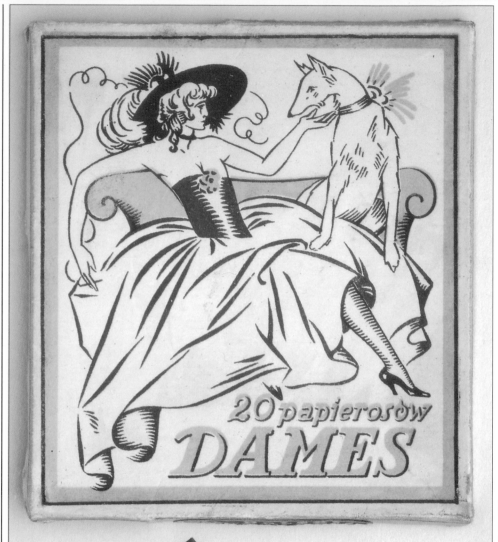

1

2

3

This Cogetama Cigar tin (*3*) was probably designed in the late 1930s.

Du Maurier Cigarettes (*4*) were relaunched by the London tobacco firm of Peter Jackson in 1929. This cigarette, the first to have a filter tip and with an elegant Art Deco box was targeted mainly at female smokers.

4

The artist M.Ponty created the pack design for the French state-owned cigarette brand, Gitanes (**5**), in 1930. The pack won many design awards in international exhibitions during the decade. The image remains virtually unchanged to this day.

Player's Empire Cigarettes (**6**) were issued at the time of the British Empire Exhibition held at Wembley in 1924 and 1925. The symbol of the Exhibition was the lion, created by F. C. Herrick.

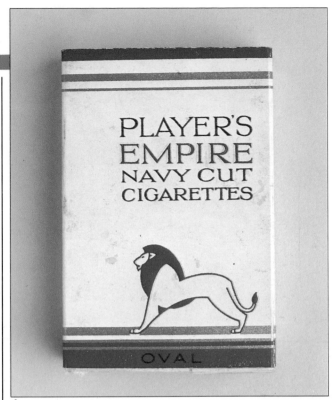

6

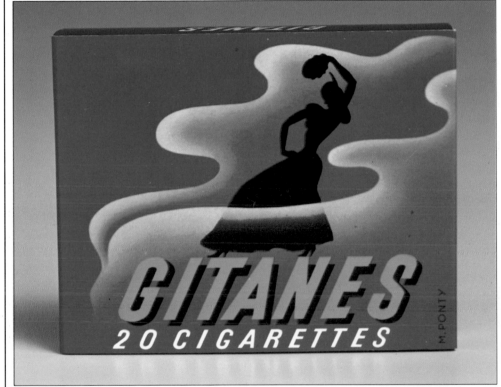

5

Captain Swift Cigars (**7**) came in an unusual cylindrical box – most cigars of any length were packed in flat wooden boxes. This design, from about 1930, reflects the climate of Havana, Cuba.

The German pack for Overstolz Cigarettes (**8**) from the late 1920s was made from aluminium, with an embossed design. It contained oval cigarettes.

7

8

BEVERAGES

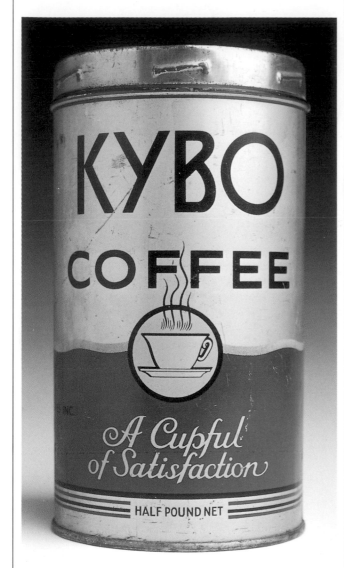

The American Ecco Coffee tin (2) dates from about 1930.

Kybo Coffee (1) was produced by First National Stores of Boston, Massachusetts around 1935.

Nescao (3) uses a French interpretation of an image for coffee (in this case a cardboard dummy display, c 1935).

In the early 1930s the Swiss based firm of Nestlé started the search for a satifactory form of soluble coffee. By the end of the decade they had solved the process of drying coffee extract and the powdered coffee was then packed in airtight tins. Nescafé (4) was launched on to the British market in 1939. World War II held back sales, but during the 1950s "instant coffee", as it was becoming known, started to take off.

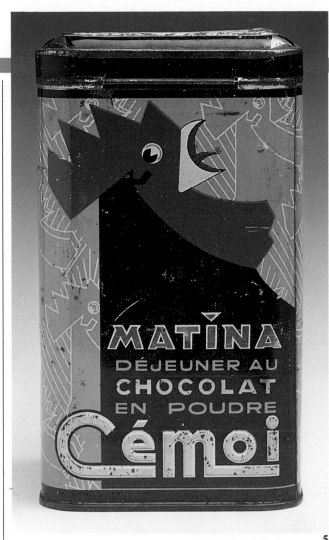

This Matina drinking chocolate tin (**5**), c 1930, from France, was specifically created so that it could be used as a storage container after the contents had been finished. On the reverse side the design is muted except for the word "Pâtes" noodles; other tins would have been reserved for flour, tea, sugar and so on.

Cadbury's Cup Chocolate (**6**) was launched in Britain around 1930 inspired by early Victorian scenes.

5   6

White House Tea (**7**) was produced by the Dwinell-Wight Company, c 1925. The image of the White House, the United States presidential residence since 1800, has been used on various occasions by American companies.

College Girl Tea (**8**) was distributed by Jenkinson-Bode of Illinois. The design for this carton probably dates from the mid-1920s.

7   **8**

MEDICINAL

Breethem (**1**), little tablets to sweeten the breath, came in a small tin which could be kept in a handbag. The American design dates from 1931.

Aspirin was produced by the German firm, Bayer. At the outset of World War I in 1914 the supply of Aspirin was cut off.

During the next year an Australian chemist, George Nicholas, created his own formula for Aspirin, and took over the trade name. However, there was still the association with the former German product, and so in 1917 the name was changed to Aspro (**2**).

The aptly named Zig-Zag (**3**) cotton wool or wadding was made by the French firm, Ruby. The way the wadding was packed enabled amounts unused to remain untouched. It was therefore economical, "one takes out each time just the necessary quantity". The pack design was based on a drawing by the artist, Loth, and first used in 1937.

This small tin of Throat Perles dating from the early 1930s (**4**), was made by Carter & Sons of Sheffield, England.

1

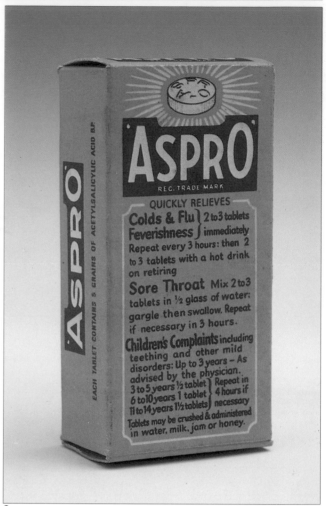

2

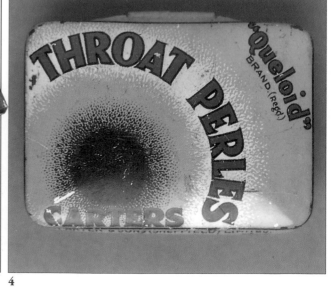

3

4

PET CARE

5

In 1892 the firm of Bob Martin (**5**) began by building up a range of pet-care products. In 1934 they decided to update their packaging into a uniform range. At first symbolic dogs and cats were considered, but these were dropped in favour of realistic illustrations. The redesign was carried out by Bob Martin's advertising department, after the submissions of packaging specialists had been rejected. The new range was launched in 1935.

AUTOMOTIVE

1

2

Oilzum Oil was created in 1905 by the firm of White & Bagley, Massachusetts, as a lubricant for automobiles. Many of the early racing cars used it, the brand being publicized by the race winners. Oilzum was the oil used in racing driver Fred Marriott's car in 1906 when he broke the world land speed record of 127 miles per hour. The Oilzum Kid character appeared on the early cans, but was replaced in 1910 by the rugged driver on this 1920s can of Oilzum Tar Remover (*1*).

In 1897 Marcus Samuel took over his father's world-wide trading business, which had started in a London shop in 1820. Since the original shop had become widely known for selling shells, the new enterprise was called the Shell Transport and Trading Company, and the shell symbol has been used since 1904. In 1906 Shell merged with Royal Dutch and in the 1930s a trading agreement with British Petroleum was reached making Shell Motor Oil (*3*) the most widely distributed in Britain.

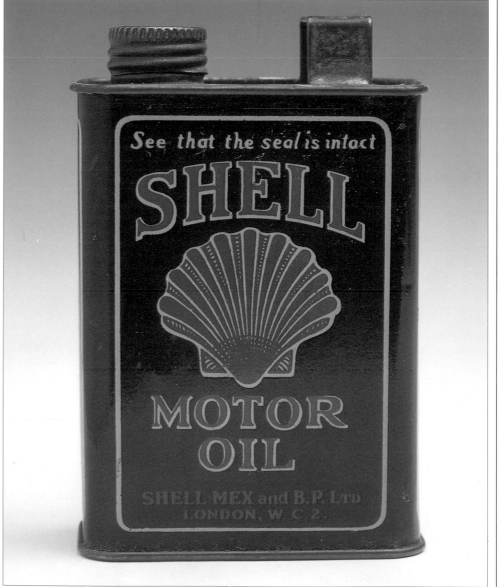

3

Johnson's Wax had been advertised as a polish for automobiles as well as for floors and linoleum. As the number of cars increased during the 1920s, a specialist polish was introduced, Johnson's Cleaner for Automobiles (*2*).

4

Another firm to catch on to the growing car-care market was that of the Simoniz (**4**) Company of Chicago. Their catchphrase, "always Simoniz a new car", helped to encourage use of their product from the moment of delivery. This tin dates from around 1930.

In Britain the firm of Reckitt & Sons were searching for new brands to add to their range of household products. In 1927 they launched the aptly named Karpol (**5**), an oil-polish finish for motor cars that could be applied even over mud and grime without scratching the surface. The Karpol tin, with its zigzag stripes, had a grooved cap into which a coin could be inserted to twist it off (The idea was patented in 1921.)

In the late 1940s many designs for Reckitt's brands, including Karpol, were updated. In this instance, the familiar 30s' poster image of two striding men was used (**6**).

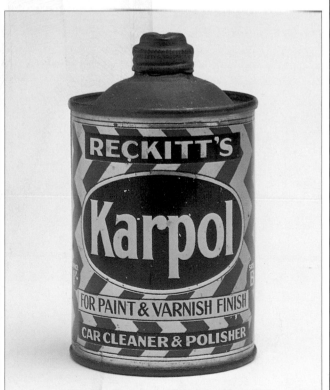

5

6

CLEAN AND SHINE

1

Samuel Johnson started a business in Wisconsin in 1886 as a flooring contractor. Many of his customers wanted to know how to take care of their new floors, and sought advice from Johnson. He began to study waxes and then formulated his own to sell as a sideline. In 1906 the firm became known as S. C. Johnson & Son and were trading from London by 1914. This British tin of Johnson's Wax Polish (**1**) dates from the mid 1930s.

Nathaniel Fairbank had first been a grain agent during the 1870s and then started a lard-rendering plant in Chicago. In the mid-1880s he launched Gold Dust Washing Powder. The trade mark of the black twins is said to have originated from a cartoon in the British magazine, Punch, which showed two black children washing in a tub – the caption read, "Warranted to wash clean and not fade". This pack (**3**), modified from the original, dates from around 1930, not long before the product disappeared from the market.

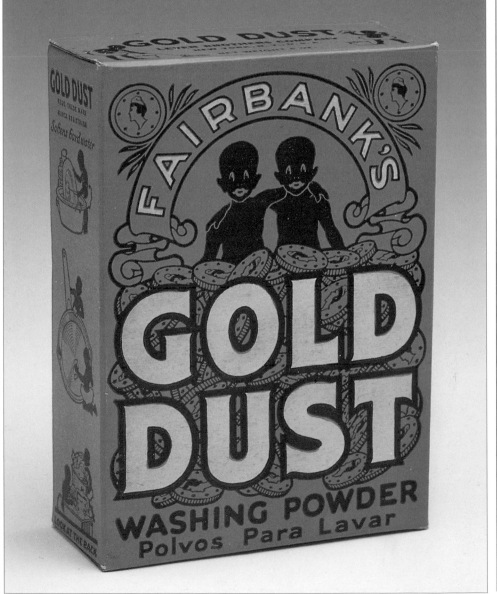

2

Polyshine (**2**), "the quality shoe polish", was the product of the Rochester Polish Corporation, New York State, c 1930.

3

Following the success of Sunlight soap, William Lever decided to launch a brand of soap already sliced into flakes. Initially called Sunlight Flakes, it was reintroduced the following year as Lux. As a product for washing more delicate clothes, including woollens, it took time to establish itself. It was first marketed in a carton with a busy design, but the image was overhauled in the late 1920s to make the name and pack design much bolder. Lux (**4**) was withdrawn in Britain during World War II owing to lack of raw materials, but returned in 1947.

5

Gustavus Swift founded his meat business in 1868 and moved to Chicago in 1876. Swift & Company was incorporated in 1885. They later began to manufacture household products like soap and borax, in addition to distributing meat, which was now being refrigerated. Sunbrite Cleanser (**5**), one of many scouring powders on the market in the 1920s, was manufactured by Swift & Company along with Quick Arrow Soap Flakes.

4

DETERGENTS AND SOAPS

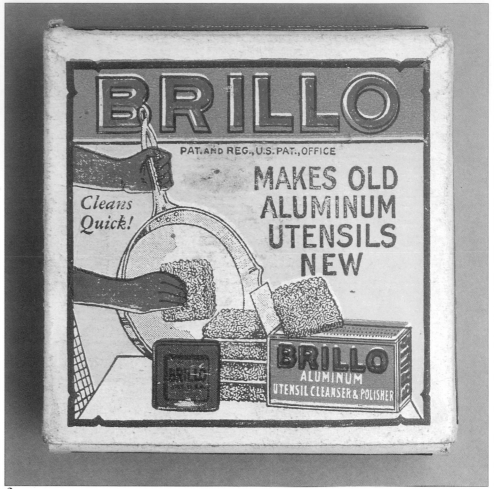

1

A Frenchman discovered how to add bleach to soap and thus invented the first household detergent. It was manufactured by a German firm, Henkel & Co, from 1907 and called Persil. The pack here (**1**) dates from around 1930.

2

3

Fairy Soap (**3**), a block soap made with olive oil, was manufactured by the British firm, Thomas Hedley & Co of Newcastle upon Tyne. Introduced in 1926, the pack was dominated by the bold lettering of the brand name. On each side were illustrations of ways the product could be used.

Brillo was launched in 1913 as a clean up agent and polish for a variety of surfaces from enamelware to golf clubs. The Brillo steel wool pad was rubbed on the Brillo soap to produce a lather. In the words of the Brillo Manufacturing Co of New York, "No effort, no waste, no rags, no powder". In the late 1920s Brillo was introduced to Britain, and sample packs (**2**) were given away; the full-size carton can be seen in the illustration.

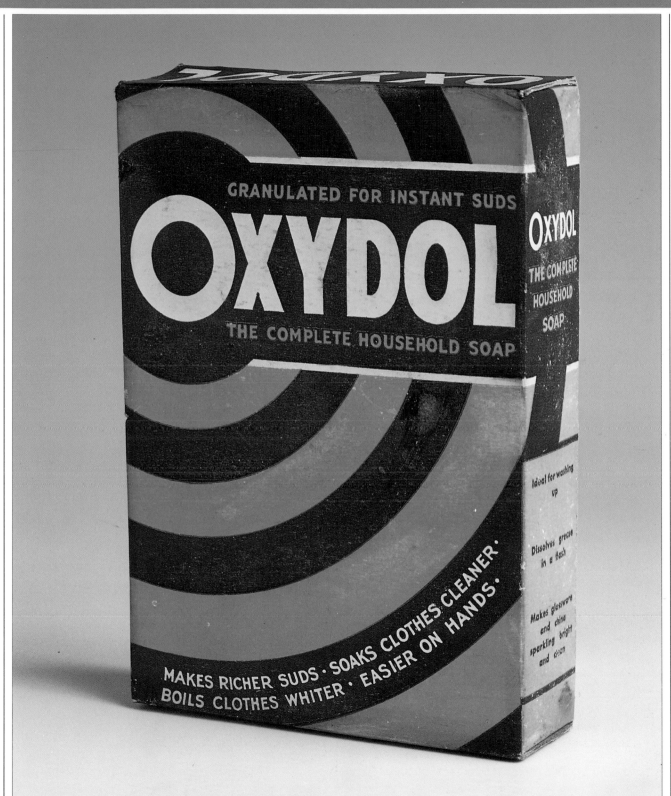

Oxydol was another detergent washing powder, this time launched by Procter & Gamble in the USA in 1927. When they acquired Thomas Hedley in 1930, Oxydol (*4*) was introduced to Britain the following year using the same concentric ring pack design.

Using a descriptive brand name, Rubon (*5*), the manufacturers added a large picture of women polishing the floor and furniture on the front of the tin. The domestic scene puts the date of the tin in the early 1930s.

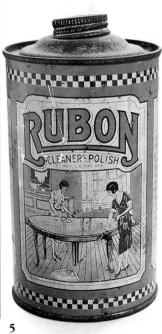

4   5

TOILETRIES

The leading brand of toilet soap in the USA was Palmolive (*1*) in its textured paper wrapper with a black paper band. It was first sold in Britain in 1923. Lux Toilet Soap (*3*) by Lever Brothers was launched first in America in 1925 and three years later in Britain. Although Lifebuoy Toilet Soap had been launched in both the USA and Britain around 1905, in Britain the brand did not succeed. But in 1933 Lifebuoy (*2*) was relaunched successfully in the bright red pack which was also used in America.

1

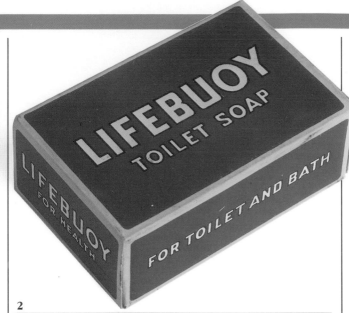

2

4

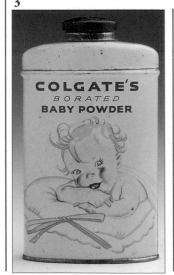

3

The firm of Johnson & Johnson was formed by three brothers in 1886. They had been in the medicine business before and continued to produce a variety of proprietary brands. An early addition was antiseptic surgical dressings. Around 1900 baby care products, such as antiseptic powder, started to be made. The Johnson's baby products (*4*) here date from the 1930s.

By the 1930s a more stylish baby appeared on the front of the Colgate Baby Powder tin (*5*).

5

6

POUR
BLONDES

SHAMPOOING
DOSE POUR UN
LITRE D'EAU

Soir de Paris

BOURJOIS

PARIS   FRANCE

7

Max Factor, a Polish immigrant, set up his perfume and make-up business in Missouri in 1909. Soon after he moved to Hollywood and his products became associated with glamorous film stars. In 1935 a branch was opened at Bond Street, London. The Max Factor products here (6) date from this time.

Soir de Paris (7) was the name given to a range of toiletries brought out by Bourjois of Paris in the 1930s.

8

Liggett's of New York produced this talc (8) around 1935.

The British cosmetic firm Dubarry launched this talcum powder (9) around 1930. It was described as "the after-bath delight, imparting cool, satin-like comfort to the skin".

9

115

FOR THE FACE

In the 1920s it was still considered undesirable for women to look sunburned. One product created to prevent "unsightly" freckles was Metamorphosa Cream (1), manufactured by Aktiengesellschaft of Berlin. The Art Deco style label of about 1925 shows the sun's rays beating down, but avoiding the lady's face.

1

2

Lasheen (3) was made by Bash & Co of London. The box, dating from the mid-1930s, contained an outfit for applying Lasheen to the eye lashes. Also enclosed was a booklet, "How to make your eyes expressive" by the film star Helene Costello. The booklet tells how Lasheen had been used mainly by actresses and the preparation had originally come from Egypt. "It is the girl with long, sweeping lashes who attracts the most attention in a crowd."

Givryl (4), c. 1930, was a French tooth powder to make teeth white.

3

4

The Butywave Company of London made a range of hair-care preparations, at a time in the 1920s when the style for women's hair was the cropped or "bobbed" look, often crimped into waves. This Butywave Shampoo tin (2) is a Norwegian one from the early 1930s.

Amami Cuticle Remover (5) ("also excellent for removing stains from under the nails") was one of the many toiletry preparations produced by Pritchard & Constance of London during the 1930s. Their most widely advertised products were shampoo powders, promoted with the slogan

"Friday night is Amami night". On many products the company gave a phonetic way to pronounce the brand name, Ah-Mah-Me.

5   6

There is no doubt that this box by L. T. Piver (7) of Paris contains make-up for eye lashes. Inside there is a mirror fitted into the lid. This outfit, designed to be carried in a handbag, dates from around 1930.

The image of a harlequin was popular during the 1920s and fitted in well with the design principles of the Art Deco movement. The Parisian cosmetic firm of Tokalon (6) produced this face powder in the early 1920s. The powdered face and painted lips are complete with beauty spot.

7

FACE POWDERS

The design for the Houbigant face powder box (*1*) from Paris was created around 1925. The powder came in nine different shades.

Outdoor Girl cosmetics made by Crystal Products of London (*2*) were typified by a female golfer in 1930s style, as used on the face powder pack shown here.

1

2

3

This box of Coty Airspun Face Powder (*3*) was thought to have been designed by the French artist, Léon Bakst, in the mid-1920s, but is considered to be the work of his friend, René Lalique, primarily noted for his glass sculpture.

The American Kissproof Powder (*4*), "lasts hours longer than ordinary powder", came in a purse-sized tin, decked out in full Art Deco style. The Kissproof trade mark was registered in 1926, but this design may have been introduced a few years later.

4

The pack design for Three Flowers Face Powder (**5**) by Richard Hudnut of New York dates from around 1920. On the Travelette box the lady holds a jar rather than a powder puff.

5

Rodoll Face Powder (**6**) was produced by P. Giraud of Paris in the 1930s using a stylized flower motif.

Tangee Face Powder (**7**) was made by Luft-Tangee Ltd of London. This sample tin, c 1930, claimed that the powder produced "a soft underglow, making your skin look younger, fresher, more natural. No dull, lifeless mask or 'powdery' look".

6

7

SHAVING PRODUCTS

Colgate's Shaving Stick (**1**) of the mid-1920s was sold not only in its own carton, but also in a "handsome nickled box".

In the 1930s many of the Mennen range of products were redesigned. Mennen Talcum for men (**2**), in a neutral tint that "won't show on your face", took on distinctive stripes and reduced the formerly large Mennen profile down to an insignificant detail.

1

2

The new designs for the Culmak Shaving Brush (**3**) were created by Milner Gray in 1934. The firm of Culmak had been established since 1800 in Suffolk, England.

Aqua Velva after shave (**4**) was produced by the J. B. Williams Company of Glastonbury, Connecticut. This bottle with its Bakelite top was made for the French market, with the instructions printed on the reverse of the front label.

3

4

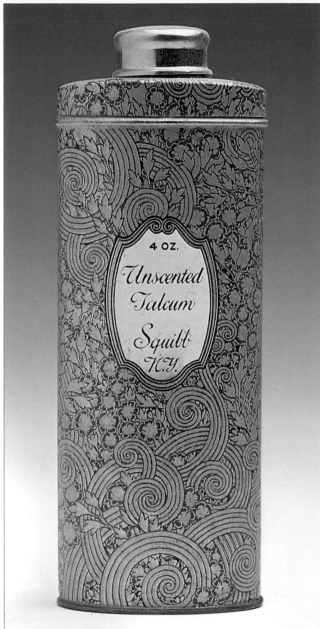

5

Dr Edward Squibb was director of the Naval Hospital, Brooklyn. In 1858 he started a business providing drugs to the armed forces. The firm was sold in 1905 and the business was extended into toiletry products. By 1910 a trade mark had been

devised: three pillars – uniformity, purity and efficiency – which supported a stone beam, reliability. This tin of unscented talcum (**5**) with its intricate design dates from around 1930.

7

In 1932 Blue Gillette Blades (**7**) were introduced, a photographic likeness of the founder displayed on each pack.

8

The pack of Ile de France (**8**), from the late 1930s, contained five razor blades, a number accepted as standard around the world.

The soapmaker B J Johnson set up his business in Milwaukee in 1898. The name Palmolive was derived from the two ingredients he put into his soap. In 1928 Colgate merged with Palmolive, which had already merged with the firm of Peet, thus becoming the Colgate-Palmolive-Peet Company. The Palmolive shaving products (**6**) here date from the 1930s.

Most packs have some sort of
identification, but this one has only the
brand name – Night Club. From the style
of the picture, it may be assumed that this
talcum powder tin was designed in the
late 1940s for the Carribean market.

CHAPTER FOUR
SELF-SERVICE REVOLUTION
1940-1959

INTRODUCTION

In Britain during World War II economies were made in packaging materials. Appeals were printed on packs to "save this carton for waste paper collection". The amount of ink used was reduced as can be seen on the Rinso pack (1) and special wartime directions for using Rinso were added on the back.

Two aerosols of the late 1950s were Dry Fly (2) by Proprietary Pressure Packages Ltd of Birmingham, England, and Colgate's insect killer, Kan-Kil.

Twenty-one years after World War I had ended, Europe was once again embroiled in war. Subservient to the demands of World War II, the 1940s saw a period of austerity. Food rationing was introduced. Natural resources were limited and the use of packaging materials severely restricted, particularly in Europe. In Britain, every possible saving on raw materials was made – paper labels on cans and bottles were reduced in size, products previously sold in tins were packed in cardboard boxes, corks replaced metal bottle tops and the overwrapping of products disappeared. Some items began to be sold without wrapping at all. Chocolate bars, when they were available, lost their silver foil and for a time their paper wrappers were replaced by thin transparent ones. As the source of materials dried up, the quality of cardboard and tin deteriorated; printing ink was used sparingly with the result that many familiar pack designs were limited to a single, often weakened colour.

2

Both during and after World War II, supplies of foodstuffs were sent to Britain from the USA and the British Empire. Much of this was canned; particularly memorable to survivors of blitzed Britain was the quantity of powdered milk and dried eggs imported to supplement the meagre wartime diet.

It was in the early 1940s that a new type of product dispenser first appeared in the USA – the aerosol canister. It had the advantages that it sprayed its liquid contents in a jet of fine mist, and both direction and amount could be easily controlled. The aerosol worked by using gas under pressure to force its contents through a valve. Such complicated packaging was costly, but this factor could be offset against the economy and ease with which the product could be used. The Americans first used aerosols during World War II to dispense pesticides in the Pacific campaign. By the 1950s aerosols were in domestic use for furniture polish, air fresheners, shaving cream, hair-spray and fly-spray.

Silk-screen printing had never made much impact in the packaging world. The process had its origins in the early nineteenth century, although the first patent was not taken out in Britain until 1907. Mainly used for printing images on to a variety of flat surfaces, silk-screen printing had also been used in the 1930s for printing single-colour

1

3

rather than asking for it, the emphasis of the pack design centred on one of instant recognition. Established brands had to stress their most recognizable points – strengthening the familiar colour, making the brand name bolder and emphasizing any motif or logo. Competition on the shelf meant that each brand had to stand out from its neighbour and sell itself, or die. Further incentives for impulse buying, especially at the check-out queue, relied heavily on the artfulness of the package designer.

With this supermarket era came a new aggression in marketing. Money-off coupons, on-pack competitions and offers, "flash" packs and price reductions were all used extensively and with increasing vigour during the 1950s and have continued to be.

On top of all this came the television commercial. The popularity of television grew rapidly during the 1950s. In America, ownership increased from three million to fifty million sets during the decade, and although the habit was slower to catch on in Europe, it had

words and images on to milk bottles. (There had earlier been a process for labelling bottles by means of a transfer.) In the late 1940s improvements were made in America and Britain in the application of "permanent" labels to glass containers using vitreous enamels with the silk-screen process. The advantage of this technique was that recycled bottles retained their printed image even after many return trips to the factory, eliminating the need to clean off and re-apply paper labels. This was particularly useful for beer bottles and, later, for bottles of soft drinks.

By 1950, the vast majority of goods sold in the grocer's store were pre-packaged. Even biscuits, which in the 1930s had predominantly been sold loose in Britain, were being wrapped individually in ½lb packs. The financial logic of the self-service store became clear. In America self-service stores were rapidly taking over from traditional grocers (by 1965 about 95 per cent of the grocery trade was self-service), while in Britain there was a more gradual conversion during the 1950s and slightly later on the Continent.

A whole new approach to packaging design was needed to cater for the self-service store. During the 1930s there had been much rationalization of style, resulting in a clearer, less complicated formula. With the customer now recognizing and reaching for the product on a shelf,

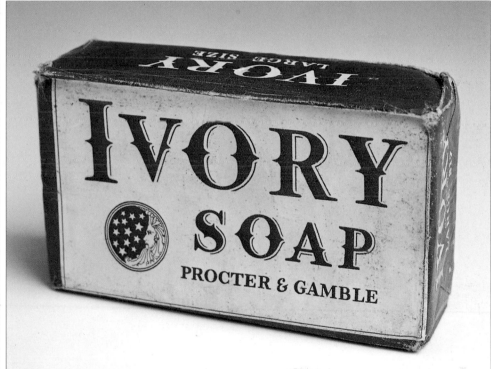

4

The black pepper tin (3) has a design typical of 1940s style. America's chain store, Grand Union offered a "delivery to your door" service.

Procter & Gamble's Ivory Soap had been launched in 1879. In 1940 the wrapper (4) was again cleaned up to give it a fresh start. The moon and stars trade mark, in use since the 1850s, was originally retained, but removed finally in 1985 after rumours connected the

mark with devil worship. The company first started in 1837 when William Procter and James Gamble made candles and soap in Cincinnati, Ohio.

INTRODUCTION

an impact on all levels of society as the realities of the outside world moved into the privacy of the home. This new form of in-home entertainment gave added momentum to the development of instant products. Something which could be prepared easily and quickly would give more time for leisure activities and for watching television; more television viewing meant that the commercials had greater influence, and in turn the new instant products (coffee or mashed potato, for example) had greater sales. The ultimate TV dinner was not far away – a complete ready meal, frozen and already laid out in place on a disposable foil tray, cooked in under half-an-hour.

The method of freezing food to preserve it for long periods of time had been known for a long time, but the inspiration to develop it into units small enough for retailing belonged to Clarence Birdseye. When fur trading in Labrador he had found that frozen fish and meat were still fresh when thawed out months later. He cleverly developed this principle for commercial use and in 1923 founded Birdseye Seafoods in New York. Fish had been frozen commercially for many years previously, but Birdseye's contribution was to freeze a whole range of perishable foods and sell them in retail packages to those retailers who had freezer cabinets.

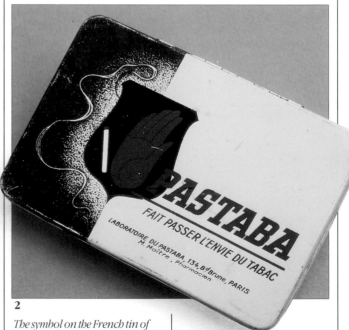

1

Since the 1870s a great area of California was developed for growing fruit and vegetables. Labels were soon being used to distinguish one orchard's produce from the next, and in time the pictorial labels became decorative and colourful. By the 1940s the heyday of the fruit crate label had passed. The example here (1) was a stock label designed in 1943, ready with its own brand name, Blushing.

2

The symbol on the French tin of Pastaba (2) indicates clearly the purpose of the pastilles – they helped smokers to resist cigarettes. The tin probably dates from the mid-1950s when the connection between smoking and health had started to come to public attention.

3

The Automatic Cone Company of Cambridge, Massachusetts, had produced ice cream cones since 1903. In the mid-1940s the "new modern pleasure for home service" were Sonny Sugar Cones (3). Customers had the chance to win a prize if they tore off the end flap and mailed it to the manufacturer, but no details of the prize were given.

The two qualities of plastic, flexibility and ease of moulding, were successfully combined in this container for Jif (4). Containing real lemon juice, it was designed by W.A.G. Pugh and launched by Colman's of Norwich in 1956.

In 1928, the rights to Birdseye's operation were acquired by the Postum Company (later General Foods), who must have seen the potential of the business. By 1931 they were supplying over 5,000 retail outlets and, by the end of the decade, it was *the* business for any entrepreneur with a bank loan to move into.

In Britain there had been a limited quantity of frozen fish and peas available as early as 1930, but it was not until Smedley's began freezing vegetables and fruit in 1936 that a proper business developed. By 1939 there were some 500 retail freezer cabinets in use, but the war put a stop to further expansion, and the frozen food market did not begin to expand until the 1950s. Much of this expansion was due to the efforts of Birds Eye. (General Foods negotiated a deal with Unilever in 1943, to expand Birds Eye frozen foods outside the United States).

The packaging of frozen foods was usually in the form of a carton with a waxed paper wrapper, the contents heat-sealed in a moisture-vapour-proofed film. In the early 1950s two further methods were tried: a composite container consisting of a fibreboard body and metal ends and a semi-rigid wrapper of laminated foil and film. Both had only limited success because by the 1960s the virtues of polythene bags had become apparent.

Advances in the use of plastics had been made slowly throughout the century, but by the 1950s the flexible plastic package had fully arrived. To the designer, the moulded pack, which gave endless possibilities for shape, offered great scope for innovation. To the consumer, the squeezable plastic bottle gave the most satisfaction; it could be used for a wide range of products from talcum powders through to the newly arrived washing-up liquids.

4

It was the combination of shape and squeezability that made such a success of the Jif "squeeze lemon". The plastic lemon captured the imagination of the British public. Apart from its novelty value, it was able to dispense a controlled amount of juice without mess or waste and preserved the remainder for future use.

The squeezable plastic tube followed the plastic bottle. Initially it did not compete with the metal tubes, but during the 1980s most of the vast toothpaste market moved into plastic tubes. Cellophane film was increasingly used, and was developed particularly for the purpose of wrapping biscuits. The much-maligned "blister" pack was also a product of this period. Nothing could stop "progress", and for the next generation the progress in plastics was to be in polythene.

TOBACCO PRODUCTS

Since 1890 the tobacco firm of Rothman sold cigarettes from London's Pall Mall. De Luxe Pall Mall cigarettes (**1**) were launched in 1947 with a design by Eric Fraser.

The updated Velvet Tobacco (**2**) tin of the early 1940s incorporated a design based on the shape of an American army badge. The badge emblem showed a cigarette as well

as a pipe to emphasize the point that the tobacco could be used for roll-your-own cigarettes. The base of the tin had a rough surface on which matches could be struck.

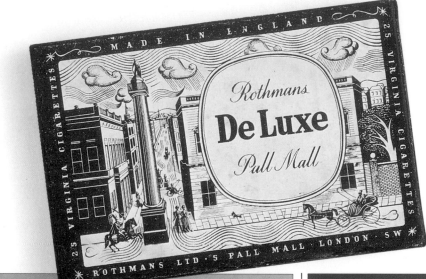

1

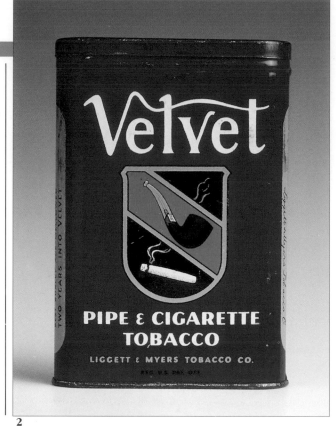

2

3

4

Bus Cigarettes (**3**) were manufactured by the National Tobacco Company of India. The graphics date from the period following India's Independence in 1947.

Philip Morris opened a tobacconist in London's Bond Street in 1847. The American company was incorporated in 1902. Marlboro was originally launched as a cigarette for ladies in 1925. There was even a version with a "beauty tip" – a red-coloured end to conceal lipstick marks. In 1955 Marlboro was relaunched in a red and white pack with a flip-top box (**4**), an invention of Molins Machine Company of London.

WARTIME FOODS

1

2

Two of the most memorable products of wartime Britain were dried eggs and dried milk (1). Many of these products arrived from America, accompanied by appropriate graphics.

The paper shortage in Britain during and immediately after World War II necessitated economies in packaging. No longer were cans or tins overwrapped in paper, nor were bottles and jars individually boxed for added protection. Further economies were made by reducing the width of labels as can be seen here (2). To save on ink costs much of the colour was also removed.

FOOD VARIETY

Having launched Corn Flakes in 1906, Will Kellogg continued to create new breakfast cereals such as Rice Krispies, launched in 1928. The Snap, Crackle and Pop characters arrived in 1941. A British company was formed in 1924. In 1938 Kellogg's innovated the variety pack (**1**) containing ten small versions of six different Kellogg cereals. Handy individual packs, each with enough for a single serving proved popular. The variety pack arrived in Britain in 1958.

The race to introduce sugar coated cereals was won by Kellogg's with their Sugar Frosted Flakes, introduced in the USA with the character, Tony the Tiger, in 1952, and in Britain in 1954 (**3**).

2

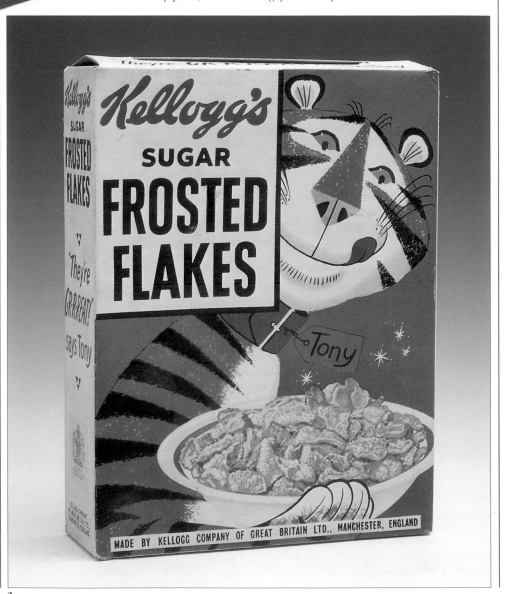

3

It was in 1840 that two Scottish muslin firms, William Brown and John Polson joined together in Paisley to form a new company as bleachers and starchers. A powder starch was soon created from sago, and in the mid-1850s cornflour was sold for food preparation. Brown & Polson Variety Custard (**2**), c 1955, gave the family a choice of six different flavours.

4

5

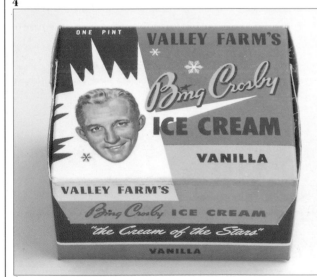

6

7

Valley Farm launched Bing Crosby Ice Cream (6) in 1953. It was manufactured under licence of Bing Crosby Ice Cream Sales Inc of Hollywood. A "pledge of quality" was given on the pack, guaranteeing the high standard of ingredients. At this time Crosby was at the height of his popularity; he had recently made the film "Road to Bali" and had also published his memoirs Call Me Lucky.

Prima Salted Peanuts (4), the product of the International Nut Co of London, were opened by twisting off a band of the metal side with the key provided. During the 1950s this was a popular way of sealing airtight tins.

As sales of frozen foods picked up during the 1950s, their packs became more appealing. This Birds Eye packet (5) for the British market dates from around 1955.

By the end of the 1950s, images that had been photographed rather than drawn by hand had become common. It was probably thought to be more realistic – and the camera never lied. Viota Cake Mix (7) was made by Stoddart & Hunsford of London.

SWEET DRINKS

Both sides of this French pack for Menier Cacao Sucré (**1**) are shown. The flavour of this mid-1950s pack has been focused on traditional Dutch scenes. It may seem strange, but the directions for making the cocoa were printed identically on each side. No matter which side was looked at first, the directions were immediately visible.

1

Time has stood still for Eleven O'Clock Tea (**2**) as the image has remained the same since its inception – probably in the late 1940s. Grown and packed by B. Ginsberg of Cape Town, South Africa, the product was guaranteed free from caffeine, harmful alkaloids and irritating stimulants.

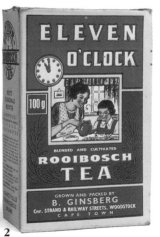

2

Maxwell House Instant Coffee (**3**) was launched in the USA in 1940 and in Britain in 1954. This late-1950s' pack saves 4d off the standard price, an early example of a money-off "flash" in the UK.

J. Lyons & Co started as caterers for the Newcastle Exhibition of 1887. Later they opened a chain of teashops in London, the first in 1894. The company started to pack tea in 1905 and during the 1920s coffee, ice cream and cakes were all sold as packaged items. This tin for Lyons French Coffee (**4**) was designed in the late 1950s.

3

4

5

Kia-Ora (the Maori name for "good health") was originally produced as a lemon squash by John Dixon of Sydney, Australia in 1896 and came to Britain in 1913. This Kia-Ora Suncrush (**5**) label was designed by Loudon Sainthill in the mid-1950s – the sun's rays strike through the ripened oranges to great effect.

6

A typical piece of graphics dating from around 1950 was that for Schweppes' Black-currant Flavour Cordial (**6**). The firm has been most closely linked with aerated drinks. It was in the 1780s in Switzerland that Jacob Schweppe invented an apparatus for carbonating drinks. In 1792 he moved from Geneva to London, where he set up a mineral water factory.

Manufactured by Nestlé's, Nesquik (**7**) was launched in Britain in 1955, following its introduction in America in the 1940s.

7

Henry Tate founded his sugar refinery in 1869. This pack for Tate & Lyle Cube Sugar (**8**) was designed in the early 1950s.

8

BOXED SWEETS

George Bassett first made sweets in 1842 in Sheffield, England. His best-known product, Liquorice All-Sorts, came about by accident in 1899 when boxes containing different varieties of liquorice were knocked over in front of a potential buyer. Bassett's Dolly Mixture (*1*) was another line, which became popular in the mid-1950s.

This dummy pack of Carsons Chocolate Orange Creams (*2*) was made for display only. At the time it was produced, in about 1955, chocolates were still being packed in their own individual paper doilies – by the end of the 1960s, pre-formed plastic trays separated the chocolates. The firm of Carsons was based in Bristol, England.

1

2

Huntley & Palmer's Mint Creme Biscuits (*3*) pack was decked out in the typical graphics of the mid 1950s.

3

The British biscuit company Peek Frean built a factory at Bedford in 1913 to manufacture chocolate biscuits and confectionery. In 1923 Meltis Ltd was registered. Meltis Savoys (**4**) were a creation of the 1950s, the contents alluringly displayed on the pack.

These chocolate liqueur cherries (**5**) were made by the Czechoslovakian firm, Rara of Modrany, in the 1950s.

4

5

6

Fruits Confits (**6**) were packed in a wooden box and manufactured by L. Tacussel of Carpentras, France. The label design probably dates from about 1950.

Town Made Assorted Chocolates (**7**), c1950, created the illusion that their chocolates were home made – by two girls using their own cooking stove. In fact they were made in a factory by the Town Made Chocolate Company of Cambridge, Massachusetts.

7

HAIR AND SKIN CARE

Nivea Cream was originally produced by the German company, Beiersdorf, in 1922. A factory was set up in England in the 1930s. Nivéa Crème Liquide (**1**) was made for the French market in the 1950s.

The stylish bottle with a plastic top for Three Flowers Brilliantine (**3**) was made for the international company, Richard Hudnut of London, New York and Paris. This example was made for the British market in the 1950s.

Eau de Gibbs (**2**) was a mouthwash made by the French branch of the London firm of D. & W. Gibbs Ltd, established around 1800.

Halo Shampoo (**4**) was manufactured by Colgate-Palmolive-Peet, New Jersey, in the late 1940s. Most shampoos had previously been made up at home by adding water to a powder from a sachet.

The pack for Jergens Lotion (**5**) was updated in 1957. The tear-drop shape and label motif was thought to have strong appeal for the female market. The manufacturing firm was the Andrew Jergens Company of Cincinnati, Ohio.

The firm of Golden Peacock Inc based their American distribution operation for cosmetics in Paris, Tennessee. The graphics for Golden Peacock Face Powder and Vanishing Crème (**6**) – a vitamin cream containing vitamins D and F – date from the 1940s.

5

6

As a souvenir of the liberation of Paris in 1944, this face powder (**7**) was produced by Leopold of Paris (the Institut de Beauté). Round the side are shown three women, each holding a flag for Great Britain, France and the United States of America.

John Woodbury of Cincinnati, Ohio, made this face powder (**8**) in the 1940s. The firm had been founded towards the end of the nineteenth century, and in 1891, like many other companies, their founder's portrait was put on the pack. It became known as "the neckless head".

7

8

TOILETRIES

During the 1950s there was a movement towards shaped novelty soaps. One such was "Fifi" (2) the French poodle made by Cussons of London.

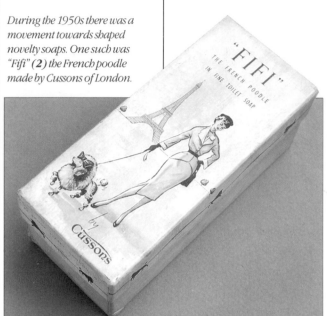

1

By the 1930s a lightly tanned skin was considered acceptable, reflecting a healthy body, and by the 1950s it had become commonplace to "take the sun" and bronze the body. A suntanned figure was now desirable and products such as the French product Tanol (1) were widely available.

2

3

Aluminium foils had first become available during the 1950s, and packs made from them had a modern sparkle. Lyril (3) was launched by Hudson & Knight (part of the British firm of Lever Brothers) in 1959, two years after Lever's Lux Toilet Soap (3) also went into a metallic pack.

Amami Lavender Bath Salts (4) had a basic design format which was carried over on to the individual sachets inside. The colour of the bath salts adds to the effect. In Britain in 1940 purchase tax was added to such products, and a notice to this effect has been stamped on the carton.

The Disney characters have been popular throughout the world since their introduction, and no more so than in Britain where Cussons launched their Disneyland Baby Powder (5) in the early 1950s.

5

The advance in the technology of plastics manufacture enabled firms to mould packs into virtually any shape they desired. It also gave them the flexibility required to squeeze or "puff" out the contents.

Two examples of plastic packs from the early 1950s – Dubarry's Puffer Talc (6) and Vinolia's Baby Powder (7) in the shape of a teddy bear – showed the possibilities of plastics in the years to come.

4

6

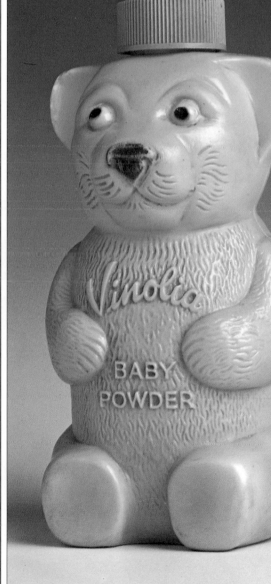

7

BATHROOM PRODUCTS

In Britain the Gibbs Dentifrice aluminium container (**1**) was redesigned in 1947. Previously the aluminium was left unpainted, with the lettering embossed. It was usual for each member of the family to have their own container and a space was provided on the base where the owner's name could be marked. The new tins were painted in different colours – red, green and blue – to enable the family to identify theirs more easily, as well as to make the pack look more cheerful.

1

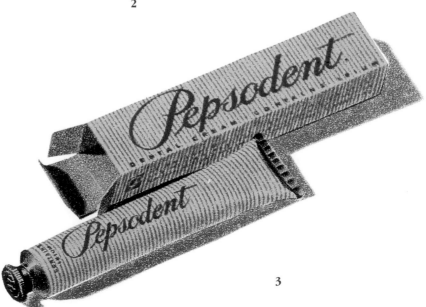

2

Pepsodent Toothpaste – an American product launched in 1915 – reached Britain in 1920. In 1936 Pepsodent was relaunched with a new ingredient, irium, which "removes the film that makes your teeth look dull – uncovers the natural brilliance of your smile". On both sides of the Atlantic the pack (**3**) was given a striped face-lift in the mid-1940s.

Gibbs S.R. Toothpaste was launched on the British market in 1934. It had a new ingredient – sodium ricinolinate, shortened to S.R. – to keep gums healthy. A new advertising campaign came in 1953 in which Gibbs S.R. (**2**) was frozen in an ice block with the slogan "tingling-fresh". When television commercials started in Britain in 1955, one for Gibbs S.R. was the first to be shown.

3

4

5

Imperial Leather was the name given by Cussons of London to the toilet soap they launched in 1938. The talcum powder followed soon after, and the example shown here (**4**) was produced in the 1950s.

From the 1930s to the 1950s, when electric razors started to make an impact, thousands of different brands of razor blade were available from anonymous manufacturers. They must have been cheap to

6

produce, as they were bought in bulk and then packaged for immediate requirements. Often they were sold from street corners or market stalls. Two from the mid-1940s were the French-made Ile de France (**5**) and the English-made Big Ben (**6**).

7

It was the International Cellucotton Products Company of Wisconsin (and later Chicago) which launched Kleenex in 1924 for use in hospitals. By the 1930s their appeal had spread to domestic use. In 1938 the Kleenex Tissues box (**7**) was redesigned in a style more frequently associated with the 1940s. The "serve-a-tissue" box also had flaps at the back which enabled it to be hung from thumb tacks.

HOUSEHOLD

Cadet Lightning Leather Dye (**1**) was made in the 1940s by the Whittemore Brothers Corporation of Cambridge, Massachusetts.

Another type of shoe dye with its own applicator was Scuffy (**2**), specially formulated for children's shoes. Promoted in the 1950s by Mickey Mouse, Scuffy came in a "non-tipping bottle" and was manufactured by K. J. Quinn & Co of Massachusetts, makers of fine leather finishes since 1880. If customers were not completely satisfied, a guarantee entitled them to a refund of the purchase price.

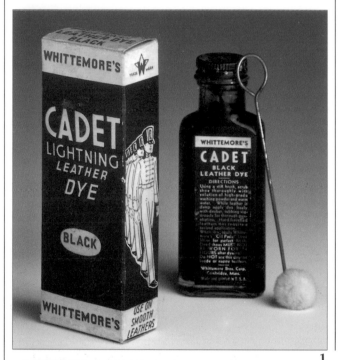

1 2

Brillo came to Britain in 1929, with a factory built in 1933. Previously in a green-dominated pack, the overall colour was changed to red in the 1950s. This economy pack of Brillo (**3**) was known as the "hotel size".

3

The brand name Sanpic had been used for a drain cleaner in the 1920s. Sanpic was relaunched as a disinfectant in 1936 by Reckitt & Colman of Hull, England. During the 1950s there were many brands of disinfectant on the market, and this label for Sanpic (**4**) was used at that time.

Trix Liquide (**6**), using the character of an elephant in overalls, was a French product, probably designed in the early 1950s and manufactured by Geigy.

Tintex was a powder dye made by the firm of Park & Tilford, New York, which was established in 1840. Tintex was also manufactured in England from the 1920s. This British pack of Tintex (**5**) dates from the 1940s.

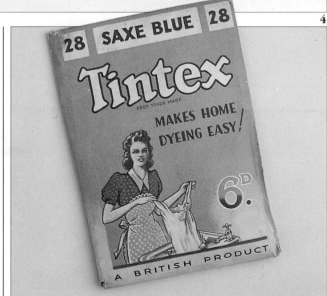

4

5

6

DETERGENTS

During the 1930s there had been extensive research into the production of synthetic or soapless detergents. In the post-World War II period, shortages of natural oils and fats stimulated research for substitutes. The British firm of Thomas Hedley (acquired by the American company Procter & Gamble in 1930) had produced a soapless powder called Dreft just before the war and this was revived in 1948.

Also in 1948 one of Lever Brother's firms, Crosfield's, launched a new powder, Wisk, onto the British market, but this was replaced by a better product called Surf (1) brought out in 1952. Surf claimed to "lather like magic" and get the whole wash really clean.

1

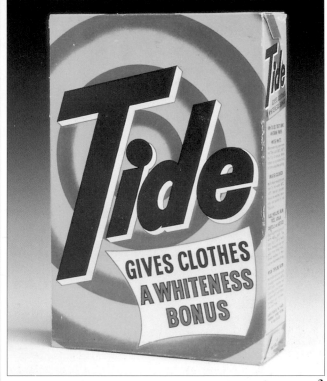

*Super Suds (**2**) were manufactured by Colgate-Palmolive-Peet in the 1940s. Said to be the only soap made specially for washing dishes, it could also be used for washing silk stockings and "dainty garments".*

*In 1946 Tide was launched in the USA by Proctor & Gamble, using a new type of organic detergent base. The British public had to wait until 1950 for Tide (**3**) to be launched. The same design was used in both countries, but while it continues to be used in the USA, the pack was completely redesigned in Britain in the mid-1960s.*

*The soapless detergent from Colgate-Palmolive was Fab (washes everything Faster And Better than soap). Launched around 1950 in the USA, Fab (**4**) arrived in Britain a few years later.*

*In 1953 Thomas Hedley brought out Daz, only to be countered by another of Lever Brothers' firms, Hudson & Knight, with their product Omo (**5**) in 1954.*

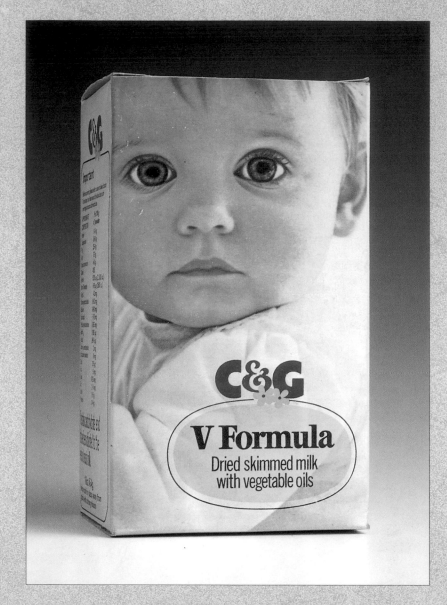

The West Surrey Central Dairy Co was
formed in England in 1882. Three years
later they had adopted trade mark was a
cow looking over a gate, and from 1908
the firm became Cow & Gate. Since then
they have concentrated on baby foods.
This appealing Cow & Gate Formula
pack dates from 1975.

CHAPTER FIVE
TECHNICAL INSPIRATION
1960-1979

INTRODUCTION

An "instant" TV dinner could be provided within 25 minutes by baking the Birds Eye Roast Beef Dinner (1) in the oven. This British product of the early 1970s put paid to the "need to worry about getting the ingredients together, or the cooking, or even the washing-up!"

The supermarket chain of Sainsbury's started in 1869 when John Sainsbury set up a dairy in Drury Lane, London. Further shops were opened and his fourth, in Croydon in 1882, was fitted out with marble counters and tiled floor. By then he was selling groceries as well and many of his products were packaged with his name on them. By the early 1960s there were over 230 branches, many of them supermarkets. The "own label" packaging (2) had gone through many phases to arrive at a clear, clean, modern style.

1

The 1960s and 1970s heralded an age of space travel and computer technology. By this time many homes in both the United States and Europe not only had colour television sets, but also fridges and often freezers as well. The kitchen frequently contained the full range of gadgetry and in the garage there was at least one car, used increasingly for picking up groceries from the supermarket, where an ever-extending range of products was becoming available. These included a variety of crisps and snacks in Cellophane bags, stacks of canned drinks and a wide choice (mainly in Europe) of dairy products such as flavoured yogurt, promoted as being healthy. Waist-conscious women aspiring to the new slimline "Twiggy" look demanded low-sugar drinks and dietary nibbles. Calorie counting had by the end of the 1970s turned into an obsession with a multitude of different diet plans and fitness programmes.

Another addition to the spice of life came from ethnic restaurants and fast food take-aways. Along with the increasing number of holidays abroad, the eyes and palates of the public were becoming aware of the culinary delights offered by other countries. In time, food manufacturers provided "foreign" meals in packets that could be experimented with at home. These came in the form of dehydrated and dried ingredients, or in a more sophisticated frozen form.

2

Eating habits, along with the fashions of the 1960s and 1970s, were changing fast. Pack designs aimed at an up-to-date look, and used photographic images, especially to show what the pack contained inside or how the product would look once served. The "own label" products of the big supermarket chains also attempted a sophisticated image, although their products were still seen as a cheaper alternative. This perception was reinforced when, at the end of the 1970s with inflation soaring, a number of chain stores produced further ranges of cheaper items that were designed to plug the "no frills to save money" concept of basic packaging.

While all these developments were taking place, technology was making further advances into the art of packaging. The public were undoubtedly aware that their familiar glass milk bottles were being replaced with an unsmashable but troublesome plastic replica or a waxed carton substitute. But they may not have been so aware of another less obvious change – the self-adhesive label that was silently revolutionizing the packaging industry.

As far back as 1935 an American from Los Angeles, Stanton Avery, had discovered the principle of an adhesive that could peel off a backing paper and then stick to another surface. Originally these sticky labels had been used chiefly in business offices. They had also had a limited packaging use, particularly for weight marking

4

During the 1960s and 1970s many products that had been packed for decades in glass bottles were switched to plastic containers. Plastic was lighter, saving on transport costs and making shopping easier. Also, plastic bottles did not break. Many household products like bleach and liquid polish were moved to plastic containers and, in the toiletries market, hand creams and shampoos as well, such as Johnson's (4) in 1970.

and price stickers (which came into their own as self-service stores developed). During the 1960s the self-adhesive label advanced to a stage where it could compete with the traditional paper label, with significant advantages over it. For instance, since the self-adhesive label came directly from a continuous backing reel it was easy to place precisely on each bottle. In addition each label was the same description, something essential for the pharmaceutical industry. A self-adhesive back could stick better to irregular surfaces and to a variety of materials. Self-adhesive labels also had the advantage that they needed no drying time.

During this period many advances were made in the use of aluminium foils. They had been in existence for some time; card milk bottle tops had been gradually phased out since 1935 and had been replaced with aluminium ones; Oxo meat extract cubes, previously packed individually in card packs, were wrapped in aluminium foil from the mid-1950s. By the 1960s, aluminium was also being used for sachets of powdered soup and coffee, to wrap cheese and butter (in a parchment and foil laminate) and for toilet soaps. When overprinted in an attractive design, the metallic foil showed through to good effect. Foil was especially useful in the area of pharmaceuticals where it provided individual casing for pills and tablets.

During the 1930s ICI developed polythene but at that

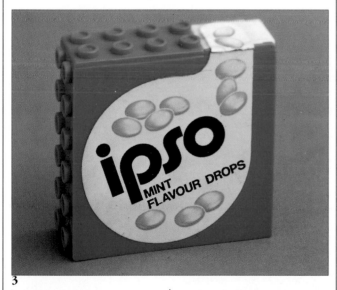

3

The inspiration for the Ipso (3) pack was the children's toy building bricks, Lego, originally made in Denmark in the 1950s. In the same way that Lego bricks could be built up, so too could the plastic Ipso packs with the added bonus that each brick contained sweets. Ipso was introduced to Britain in the late 1970s by Nicholas Laboratories Ltd.

INTRODUCTION

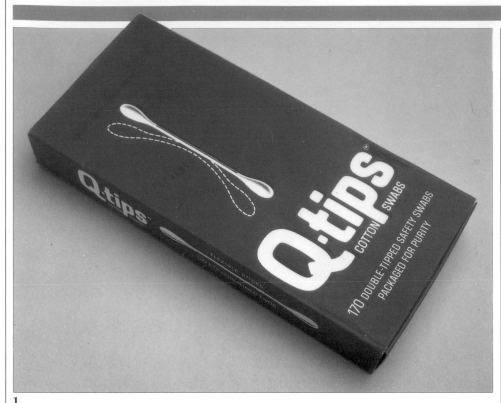

1

Q-Tips Sterilized Cotton Sticks were sold in both America and Britain during the 1950s. Bought by Chesebrough Ponds Inc and called Q-Tips Cotton Swabs, this pack (1) of the late 1970s clearly demonstrates their flexibility.

aluminium, they did not require a bottle opener and were largely an extension of the milk bottle top principle. Another type using aluminium was the external screw cap, a closure with increasing possibilities since its first use in the 1930s. By the 1960s an extension to the idea was being used in the spirits market, where tamper-proof closures were necessary. This was an aluminium closure made so that when the top was unscrewed, it split away from its locking band below; thereafter, the cap could be used in the normal way.

Apart from tamper-resistant closures, some child-proof closures were developed during the 1970s, again made from moulded plastic. They were quickly taken up by the makers of drugs and poisonous substances such as bleach.

Waxed cartons for milk had been used in the 1930s, but a Swedish firm developed an ingenious but simple four-sided carton made of paper plastic laminate that eventually led to a vast enterprise. The Tetra Pak started out in

time it was not cheap enough for packaging purposes. With the aid of their American associates, Du Pont, ICI later realized that, if ethylene was added a cheaper material, called polyethylene, could be produced. This material has been used extensively since the 1960s for a variety of packs, but in particular for plastic bottles of all shapes and sizes. (In the 1970s the use was extended to the successful polyethylene terephthalate [PET] bottles for carbonated drinks.) It was used in sheet form for the ever-expanding frozen food market, and also as a replacement for the waxed bread papers as it allowed much faster wrapping speeds. All these products could now be carried away in the polyethylene carrier bag, which had virtually replaced the paper carrier bag by the end of the 1970s.

A continual search in the history of packaging has been that for the successful closure of bottles, their subsequent opening, and their reclosure. A later requirement was for tops that were tamper-resistant. One invention that fulfilled all these requirements was the top developed by a Frenchman, Jean Strobel: a hinged snap cap moulded in one piece out of polyethylene. It was first marketed in the USA in the mid-1950s, but unsuccessfully. Then, in the early 1960s, the licence for the UK was bought by the glass manufacturers, Johnsen and Jorgensen. After modifications, the top became a highly successful closure for millions of bottles.

Other bottle closures that made their mark in the 1960s were those that could be peeled off by hand. Made of

2

In the 1960s the market for canned pet foods started to increase rapidly. One product at that time was Rival Dog Food (2)

made by Rival Packing Co of Chicago. The label carried the message, "Honestly, Rival Dog Food is wonderful".

the early 1950s as a pack primarily for milk. Formed from a paper tube, it was pinched together at regular intervals to create a pyramid-shaped container. The shape was certainly novel and the milk easy to dispense by cutting off a corner of the pack with a pair of scissors. Another advantage of the shape was that the container was difficult to knock over.

The Tetra Pak provided a number of economies both in production and distribution that gave it the edge over the existing Swedish half-litre milk bottle. For instance, twice as much milk could be loaded onto a truck. For a time in the late 1950s and the 1960s, the triangular packs were used in Britain. But the Tetra Brik, introduced in the 1960s, using a conventional rectangular shape, was preferred both by manufacturers and consumers. During the 1970s it became a popular container not only for milk, but also for soft drinks and fruit juices, replacing the heavy and breakable glass bottles.

4

3

In the 1960s the then-fashionable London clothes store, Biba, moved into a range of cosmetics and, in the early 1970s, into a variety of grocery products. Much of the *packaging was done in black with gold lettering incorporating the same image as on the Biba Instant Coffee (3).*

The other container that continued to reduce the need for glass was the can. In 1960 Coca-Cola started to sell in cans, which considerably helped to expand the consumption of Coke; and many other fizzy drink manufacturers moved over to cans, sold alongside their glass bottle counterparts.

Most beer drinkers buy more than one can at a time, so brewers banded four or six together with a card wrapper (known as a "cluster pack"). During the early 1960s also, the popularity of the "party can" grew. The "party can" contained between four and seven pints of beer, but its success was confined mainly to Europe. Americans preferred chilled beer and the cans were difficult to fit into the fridge.

The final request from beer drinkers seems to have been to find a more satisfactory way of opening the cans. The can opener was not always easy to find, particularly after the third pint. A solution – the "lift tab" – was first devised by the Aluminum Company of America in 1962. A better idea on similar lines – the rip-top or ring-pull opener – was developed by the Metal Box Company in 1967, and this finally led to total consumer acceptance of canned beer. All that bottle manufacturers could do was to introduce the lightweight non-returnable bottle.

The design for Lindt's Suvretta (4) comes from the early 1960s. Rodolphe Lindt had opened a chocolate factory in the 1880s, in Berne, Switzerland, where he developed the first fondant chocolate.

AROUND THE HOUSE

Starting out in 1898 as Boston Blacking, the firm from Leicester, England, developed a rubber latex adhesive for the shoe industry in 1931. Moving into the adhesive market, they changed their name to Bostik in 1962. Zoo Glue (**1**) was a children's paper adhesive of 1968, the plastic animals forming a zoo after use.

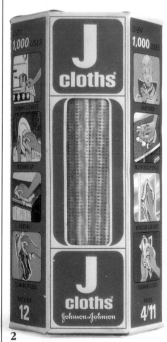

1

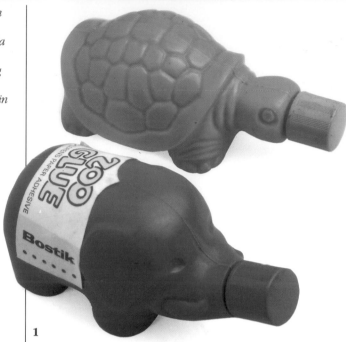

4

2

Johnson & Johnson's J Cloths (**2**) were launched in the UK in the mid 1960s. The six-sided pack has a see-through window on the front and is opened by pulling apart the interlocking tabs on top.

3

Lever Brothers launched Britain's first liquid fabric softener, Comfort (**3**), in 1967.

The liquid Ajax all-purpose cleaner was originally launched in a glass bottle in the 1950s, with the slogan "cleans like a white tornado". Then in the mid-1960s Ajax (**4**) arrived in a plastic container with the tornado embossed on the side. This example was a free sample for the UK market.

5

6

7

Kiwi Guard (5) was marketed in the late-1960s as a children's shoe polish. The liquid contents went through to a foam pad. The plastic container was dressed up as a British Guardsman with his busby top.

Silver Star Starch had been made in Victoria, Australia, since 1890. In the 1970s, this pack (6) incorporated an 1890s advertisement on the pack front.

The offer for free dress patterns has been discreetly placed amongst the roses on this American packet of Lux (7) dating from 1966.

CONVENIENCE FOODS

1

Conscious of their figures, women in the 1960s used a combination of diet and slimming products to keep down their weight. Unicliffe Ltd of Kent, England, made Limmits Crackers (**1**), "the tastiest way to slim", in the mid-1960s. At the same time another English firm, Ryvita, made Starch Reduced Wheat Crispbread (**2**), "for slimming diets in which the total intake of calories is controlled".

2

The Sanitarium Health Food Company from Auckland, New Zealand, manufactured Skippy Corn Flakes (**3**), but used the Australian kangaroo for their emblem.

Made for the festive season of 1968, this partypack of Smiths Potato Crisps (**4**) also had two recipe suggestions for food dips.

3

4

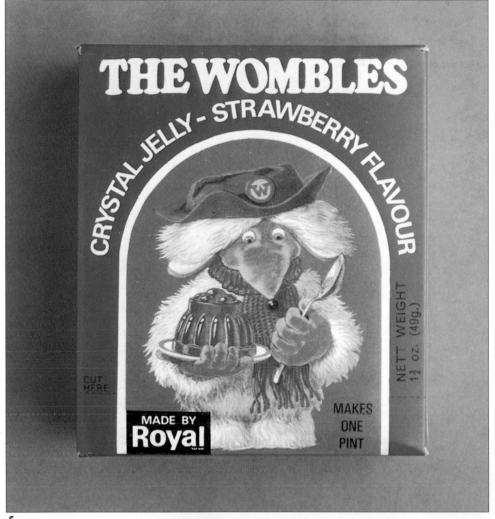

5

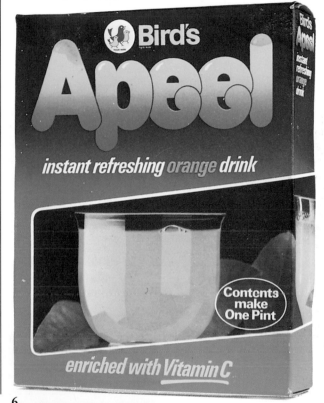

6

7

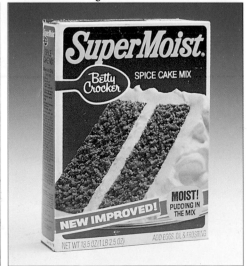

8

In 1975 the UK version of the Standard Brands product from America captured a children's television character to promote Wombles Crystal Jelly (**5**).

Made by General Foods of Banbury, England, in the 1970s, Bird's Apeel (**6**) could not succeed against the growing popularity of orange juice now available in cartons.

It was an American idea to have housewives' recipe queries answered by a fictitious character. Nabisco created Mary Baker in 1941 to be the name behind the recipes and articles published. This Mary Baker Scone Mix (**7**) dates from the early 1960s.

Betty Crocker was created in 1921 as spokeswoman for Washburn Crosby Co, Minneapolis, a flour-milling business and the forerunner of General Mills. At one time there were thirteen different people speaking as Betty Crocker on radio stations around the country. Super Moist (**8**) cake mix dates from the late 1970s.

YOGURT

In Britain yogurt had been sold in off-beat delicatessens during the 1920s and 1930s. During the 1950s the Express Dairy sold glass jars of plain and flavoured yogurt on their milk rounds. In the early 1960s, however, real fruit was added to yogurt, and the sales took off.

The leading brand was Ski Yogurt (**1**), made by an off-shoot of Express Dairies called Swiss Milk Products Ltd, based in Sussex. The first fruit they added to their Swiss recipe yogurt was bilberry, launched in 1963. The carton was made of waxed board, but had been changed to plastic by the end of the 1960s.

Raine's Dairy Products of London and Manchester had a brand called Alpine Yoghurt (**2**) made in the mid-1960s.

On the Continent yogurt was well established in the early 1960s, particularly in France with Yoplait (**3**) and Chambourcy (**4**).

By the end of the 1960s, other chilled desserts were joining the growing ranks of yogurts. T. Walls & Sons, makers of ice cream and sausages, entered this market in England with a number of ideas, including Fruit 'N Nut Brunch (**5**) and Jellymilk Dessert. Chambourcy of Paris exported Chocolat Parfait (**6**).

The Greeks (**7**) started to pack their yogurt in plastic containers around 1970.

In 1968 the Mr Men from Roger Hargreaves' children's books were used by Raines Dairy to sell yogurt to children. One example is this Mr Greedy Strawberry Yogurt (**8**).

1

2

3

4

5

6

7

8

CHILLED GOODS

The English firm of Smedley's began freezing vegetables and fruits in a limited way as early as 1936. Interrupted by World War II, they did not become established until the end of the 1950s. This pack of Smedley's Garden Peas (*1*), with its duckling pricing space, comes from the early 1960s. A drink for Superman and American children, this flavoured orange refreshment (*2*) was made by Producers Dairy Inc of California in the 1970s.

Although not entirely successful, it was intended that the Lyons Maid Strawberry Mousse (*3*) should be shaken out in the shape of the plastic mould – a racing car or a cottage. They were tried out in 1969. The London firm of Lyons Maid was formed in 1955 as the ice cream division of J. Lyons (Polar Maid was their top-selling ice cream at the time).

1

2

3

SMOKING

London Transport Shag tobacco (**4**), c 1970 was manufactured by the London firm of Roger & Durrey Ltd, established 1875. During the 1960s tobacco brands started to switch from tins to the cheaper plastic film pouches.

During the 1960s radical changes took place in the cigarette market. Filter cigarettes were becoming increasingly popular as the public realized the dangers of nicotine, king-size cigarettes started to establish themselves and, in Britain, a coupon war had begun.

When Benson & Hedges (**5**) launched their Special Filter king size brand in 1961 there were few king size cigarettes on the UK market.

4

5

6

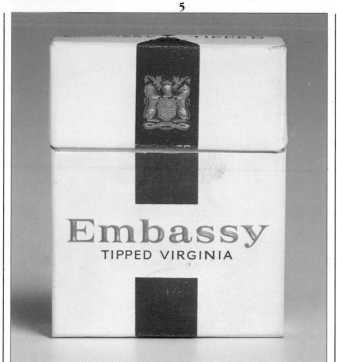

7

Many new brands were launched and many failed. New World (**6**) by Churchman's (part of the Imperial Group) was launched in 1963 but was only on sale for 18 months. A more spectacular failure was that of Wills' Strand Cigarettes in 1960.

However, Embassy Cigarettes (**7**) also launched by Wills of Bristol, was an immediate success. In 1962 Embassy arrived with gift coupons, and within three years had become Britain's leading cigarette.

BATHROOM FUN

The use of startling colours in plastic, such as that for Goya's Aqua Manda Talc (**1**) in the early 1970s, continued to encourage manufacturers to think of the possibilities for plastic containers.

Sea Witch hair colour (**2**) was made in the early 1970s by Elida.

1

2

3

Any design incorporating a woman's face had to keep the image up to the latest fashionable style. This would have been true for this British Lux toilet soap pack (**3**) of the mid 1960s.

A fun product of the 1960s, Depth Charge Sea Foam Bath (4) was made by Scott & Bowne of London, famed as manufacturers of Scott's Emulsion. Depth Charge was a creation as much in the mind as in reality. "From a South Sea lagoon we took the colour of deepest blue. We stole the fragrance from a gentle sea-breeze. And the foam from a breaking wave."

This lipstick for Elizabeth Arden (5) was inspired by the psychedelic graphics of the mid-1960s. Elizabeth Arden had opened her first salon in New York in 1910, and over the years produced a whole range of cosmetic products, building the company into a major international organisation.

5

6

Toiletries for men have always been secondary to those for women. In the 1950s and 1960s, however, many more products tempted men to think about their looks and appeal. Max Factor's Primitif Talc (6) was aimed at this growing market.

4

COVER UP

1

2

The London-based inter-national cosmetic company Goya (the name was registered in 1936) produced Head Dress (**2**) in the 1960s, the combined hairdressing cream and conditioner, perfumed with their Passport perfume.

Another 1960s brand was that made by the American company Andrew Jergen. To indicate that Jergens Face Cream (**3**) was all-purpose, for all skin types, the faces of three women, each with different hair colour, were set out on the label.

3

In the "swinging sixties", naked bodies on packs had become acceptable. All Over Softly (**1**) was a body shampoo of the late 1960s, produced by Reckitt & Colman Toiletries Division of London.

Nivea Sunfilta Cream (**4**) was produced in the early 1970s by the English firm, T. J. Smith & Nephew Ltd.

4

In Britain pink was a favourite colour of the early 1960s. Tangee Petal-Finish Make-Up (**5**) was made by Luft-Tangee of London.

Nulon was a hand-cream lotion launched in 1953 by Reckitt & Colman of Hull and London. Using this product helped to cream away roughness and dryness on elbows, knees, arms and legs. "Make a habit of using Nulon after every washing job and see the vital difference in your hands." This pack (**6**) dates from the early 1960s.

5

6

7

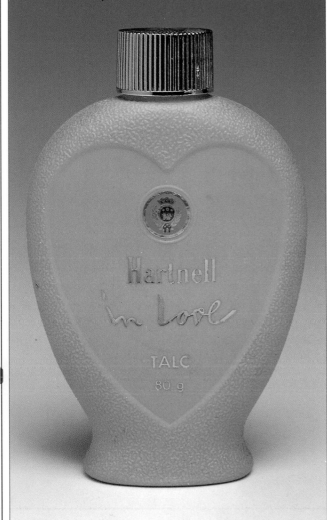

8

Another way to dispense hand-cream was from an aerosol. The English brand, Supersoft Fluffy Hand Cream (**7**), c 1965, used the novelty of the packaging dispenser to help sales, "simply press the button lightly, gently releasing enough fluffy cream ... over 150 hand beauty treatments in each aerosol".

The pink plastic container for Hartnell In Love Talc (**8**) provided a puff of powder when squeezed.

The age of plastic continues to create a variety of exciting containers. The Cosmic Raiders Bubble Bath is made by Communiqué Beauty Products Ltd; when empty the model has an end use as a child's toy.

CHAPTER SIX

DESIGNER TRENDS

1980-1989

INTRODUCTION

Ecologically safe products will be the by-word for the 1990s. These products (1) from the late 1980s lead the way, the Ecover range from Belgium "helping to maintain the purity of our rivers...are not tested on animals", CFC-free aerosols do not damage the ozone layer; and the spray gun replaces the aerosol altogether.

Instead of the traditional glass bottle, wine can now be sold in special cartons as with La Mancha, (2) imported by Tesco Stores, or in the three-litre "bag-in-box" using an air-tight plastic bag.

By 1980, most people in the Western world knew of wrapped products only; it no longer seemed strange that many of the items they bought could not be seen until the outer container had been removed. Any suspicion the consumer may have had about the content's quality had long since vanished. There was no real need any longer to add medals of excellence to the pack, although they could still be used to give a touch of glamour to the product. Neither was there still the need to offer the customer monetary rewards, as with the tightening of government regulations, the public was no longer afraid that products were adulterated. It was, however, and still is customary to add a guarantee of quality, whereby the customer may return the article "if not entirely satisfied".

Nevertheless, in the late 1970s and the 1980s, there has been a trend towards capturing the image of packs from times past, not simply for nostalgia, nor just once more to make the design stand out on the shelf, but to hark back to

2

a perceived image of purity and wholesomeness. There had been a time in Britain during the 1930s when nostalgia was for the mid-Victorian period, and fashionable ladies in crinoline dresses displayed themselves on the latest packs.

During recent years, however, there has been growing public concern on a number of levels, all affecting packaging. In its widest sense, there has been worldwide concern over environmental and ecological issues. Many of the fears relate to acid rain, chemical emissions, energy consumption and, of particular importance to the packaging industry, the effect of chlorofluorocarbon (CFC) on the ozone layer. One important source of CFC resides in the use of aerosol cans. Already there has been a trend away from the use of aerosols, and "ozone friendly" replacements claiming "this product is CFC-free" have appeared, along with hand-pump spray guns.

Another concern, particularly since the 1960s, has been the need to create packaging from materials that are biodegradable. Glass bottles can be recycled (hence the growth of "bottle banks"), paper and cartons can be recycled and are also biodegradable, but most plastic

3

In America and other countries (but not in Britain) the electric orange or yellow ring design on the Tide pack (**3**) has been retained since its launch in 1946 through to the 1980s. The new arrival has been Liquid Tide contained in an orange plastic bottle. Liquid detergents have made their impact on washday habits.

containers cannot be disposed of satisfactorily.

The problem of litter has been around for some time, and anti-litter campaigns have been mounted at intervals. From the 1960s, some packets, especially those used outdoors such as crisp wrappers, sweet packets and drink cans, added a message to the container like "please dispose of tidily". In the 1980s a further crop of socially aware messages were added, such as "children say no to strangers" (UK), and "say no to drugs" (USA). A socially successful use of packaging in the USA was when the faces of missing children were added to the sides of milk cartons. A campaign on milk cartons in the UK was aimed at preventing crime, with slogans along the lines of "lock it or lose it!".

In recent years there has been a growing awareness of the ingredients used in processed foods. Since the 1960s there has been an increasing amount of legislation that forces manufacturers to state on the pack the precise nature and proportions of the ingredients in a particular product. Public awareness of the need for better nutritional information has meant that manufacturers now also outline the quantities of protein, carbohydrate, fat and dietary fibre, for example, and often the calorie counts as well.

A hundred years ago, a great number of food products and beverages were promoted on the grounds that they were nutritional, sustaining, and brought sound, healthy sleep. It was not just that these foods would sustain good health, but that they would also be good for the nervous system. Grape-Nuts, for instance, warned that "brain workers are driven into dyspepsia, nervous prostration and various diseases because of the lack of food demanded by nature, from which to rebuild the daily loss occasioned by brain and nervous work. Grape-Nuts furnish, in a condensed form, these necessary food elements."

At the beginning of the twentieth century, there was a growing realization that vitamins were an important part of a healthy diet. By the 1930s some products started to mention that vitamins were present, such as oatmeal which contained Thiamin (vitamin B1): "oatmeal is nature's richest economical food source of Thiamin for nourishing nerves and aiding digestion. Mother's Oats make a vital contribution to your health and happiness".

It was also during the 1930s that a range of health foods

INTRODUCTION

*Most new cigarette brands launched in the late 1980s have been of the low-tar variety. To indicate this status, pack designs have been given a hygienic feel, suggested by a pure white background. The American tobacco firm Brown & Williamson launched Capri (**1**) in 1988 using this concept for their "trim and light, menthol" brand.*

*Sapporo is Japan's longest surviving brand of beer – dating from 1876. Since 1988 Sapporo Draft Beer (**2**) has been available in handy cans with a ring-pull that removes the entire lid, thus transforming the can into an effective drinking vessel.*

1

2

had been developed for slimming, especially rye crispbreads like Ryvita which had been introduced from Sweden in the mid-1920s: "Ryvita gives you snap and vim, makes you fit and keeps you slim". The slimming fashion gathered momentum during the 1960s with products like Limmits Crackers, "the world's first meal-in-a-biscuit to help you slim". More recently, it has become customary for any successful soft drink to have a low calorie equivalent, often containing only one calorie per can. A similar trend has also begun in the alcoholic drink market: fierce anti-drinking-and-driving campaigns have resulted in the widespread production of low alcohol or non-alcoholic beers and wines.

With the arrival of this second strata of health product – low-calorie food, low-alcohol drink and low-tar cigarettes – the packs for such products have developed their own style of low-key design. This usually takes the form of designs with a white background or muted colouring, or with a striped effect where alternate bands of colour let a soft background permeate a recognizable image.

The designer has had to cope with other pack additions. Since the 1960s there have been a number of additive

"scares" once the media and the public have become aware of the possible dangers. In 1968, for example, there was an outcry over the addition of monosodium glutamate (a thickening agent) in foods and the use of cyclamate (a slimming sweetener) in soft drinks. Shortly afterwards, products that might have been associated with using such dubious additives had to vigorously announce "cylamate-free" on the pack. In many ways this was nothing new to pack design. Since the 1950s, special offers, competitions, money-off and announcement flashes of the latest wonder ingredient in detergents or toothpaste had been catered for on the pack face. They were often to the detriment of the overall design, even to the extent that the brand's recognizability was substantially impaired. Only in the cigarette market has the pack's face been spared. Government health warnings were discreetly added to the side of the pack in the 1970s, and special announcements limited to the cellophane wrapper which is discarded once the pack is opened.

In the 1980s the need for such announcements continues. "No added artificial colour, flavour or preservative" is a common example along with "less salt", "low sodium" and "no saccharin." As well as nutritional information, list of ingredients, recommended daily allowance of vitamins, sell-by date, directions for use and so on, the all-important bar code has to be added. This unsightly addition on a pack can often be tucked away from view, although to begin with in the late 1970s the bar code caused problems where bottles had no label on the back. However, the advantage of the bar code system (first used in America in the early 1970s) has been to speed up the check-out tills in the supermarket and provide the customer with a detailed list of goods purchased. The system also helps the retailer to keep efficient control of the stock.

Shoppers in the 1980s have noticed many other changes. Larger pack sizes have been available for products such as detergents, and the idea of purchasing in bulk continues. More than ever before children have a range of products aimed specifically at them, from spaceship-shaped snacks to alphabet pasta shapes in tomato sauce, all wrapped up with the latest popular hero from children's television. (One market that has always been geared to the whims of children is that of pocket money confectionery.) A recent necessity has been for products to become more tamper evident, due to the possibility of radical protest groups contaminating products. The latest tamper-proof seals have been those now used on jars of jam.

It has been with jam that one of the newest developments has taken place – plastic jars which are not only shatterproof and lighter, but also squeezable. These have followed the plastic squeezable sauce and tomato ketchup bottles, which eight out of ten shoppers say they prefer. The convenience of such products is obvious and so long as the consumer is willing to pay the extra cost, which is often minimal, convenience packs will continue to be developed. At the same time, more convenience foods are being developed to satisfy the demand for frozen ready-made meals, being made more popular by the immediacy of the microwave oven (by 1990 some 60 per cent of households in the USA and Japan will have one, and 30 per cent in the UK).

Over the last hundred years, many of the time-consuming jobs that the domestic servant or the housewife did in the home have been reduced or replaced by machinery. Instant polishes give instant shine and ready-made meals can be cooked in a moment by microwave oven. Constantly there is more variety, more flavours in bigger sizes. Many of the old-fashioned or regional varieties might have disappeared had it not been for the never-ending quest to find some new tasty morsel like "plum scone cobbler" to tickle our fancy, and the pre-prepared meal which saves time on finding and preparing the ingredients buys a variety to life that might not be available otherwise.

3

4

During the 1980s the plastic revolution has continued to replace glass containers with shatterproof ones, as with Skippy Peanut Butter (3), and in the squeezable bottles for Heinz Tomato Ketchup. Both products were adapted in 1988. Fewer glass breakages have meant less mess when accidents occur, a development particularly welcomed by the supermarkets.

The use of plastic "cans" enables the contents to be seen, as with the Marks & Spencer Fruit Salad (4) which has a ring-pull top to remove the inner lid.

NOVELTIES

The Advent Calender (**1**) sold in Britain prior to Christmas 1988 is more than it appears. The card blocks were filled with sweets, and after Christmas could be used as a jigsaw to create five different pictures.

This Pot Pourri outfit (**2**) contains all the requirements to "make your own", the contents being displayed through the windows on either side of the demonstration gift. A convenient handle has been provided to encourage the customer to carry the product away.

1

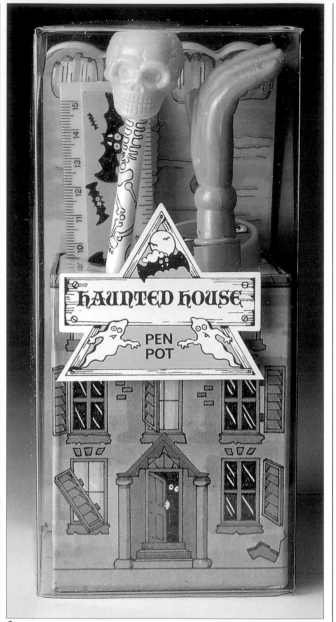

2

3

The Haunted House Pen Pot (**3**) holds a bizarre collection of stationery items and is simply packed in a clear plastic container to show off the ghoulish contents.

During the last twenty years the art of origami has been used on the cartons for British chocolate Easter eggs. The folding of cardboard has produced many extraordinary creations, as here for 1989 (**4**). The use of favourtie characters has helped sales.

1
These traditional games for children are made by House of Marbles (1). Each game is mounted on a hanging card to give it greater visibility.

2

A triangular box has been created to wrap the Philips head-phones (2), with a window that allows the customer to check the type of ear-piece.

Fisons of Ipswich, England, developed the Gro-bag in 1974. The self-contained vegetable garden has prospered as a concept. This design (3) dates from the late 1980s.

A more recent development by Fisons has been Clearcut (4). This latest instant product enables a cutting to grow from a specially prepared gel that requires no watering or feeding.

3

4

DRINKS

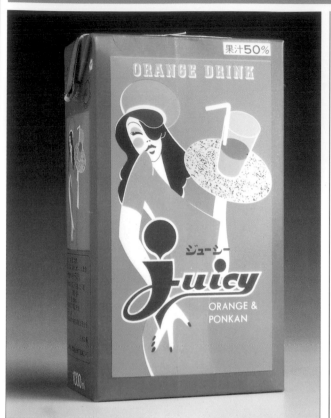

1

2

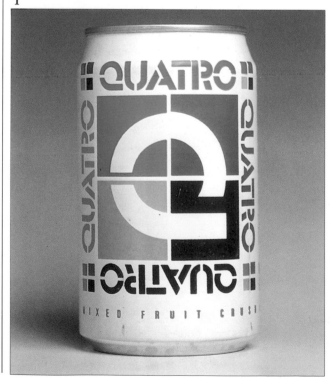

3

4

It has become acceptable for drinks to be consumed on the street, either straight from the can or through a straw from the carton. Therefore it is more important for this visible "accessory" to be tuned to the latest style. Juicy Orange Drink (**1**) comes from Japan but portrays a Western image. Coolers (**2**) from Marks & Spencer of London have a designer image for their white wine mixers. Quatro (**3**) was created by CC Soft Drinks as a drink made up from four fruits. French Perrier in a can (**4**) conveys an effervescent effect by using a design with rows of bottles.

Label designs for alcohol bottles have started to break away from their traditional look. This has been particularly evident amongst wine labels where impressionistic painters have been given carte blanche to create an image for wine. In the case of Taboo (**5**) from the Light Spirit Company, the label has a geometric design. For the English supermarket chain of Asda, Vodka has inspired an award-winning label (**6**)

5

6

FOOD

The biscuit pack *Kiss Me Licia* (**1**) was created for the Italian manufacturers Colussi of Perugia in 1986. The biscuits are moulded to the shape of the characters on the pack.

The all-plastic container for the French tomato ketchup Amora (**2**) has a nozzle that twists open before the contents can be squeezed out.

1

2

With the easing of restrictions on sales of products within the European Community, the variety of biscuits available in the shops from different countries has been increasing. In order to combat Continental competition, Huntley & Palmers (owned by Nabisco) have created a range of Continental biscuits, such as *Butter Lace* (**3**), made in France and packed in cartons that have a Continental look.

3

Appealing to the children's market, products like Titan's Meat Balls (**4**) grab the attention. The comic-strip graphics used on the labels for children's spaghetti-shape products (**5**) also appeal, particularly when linked to a TV character like Roland Rat, or the popular film Ghostbuster created the Noodle Dood character for themselves i

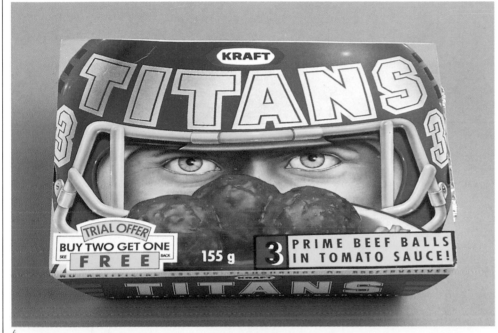

4

5

6

Distinct from the traditional label for Lea & Perrin's Sauce, the Chilli and Garlic label (**6**) has taken a modern design for this product launched by the same company in 1988.

SNACKS AND CEREALS

During the 1970s and 1980s there has been tremendous growth in the market for children's snacks, both in flavoured potato crisps and corn extruded shapes. The graphics used vary from the Muppet like Monster Munch (**1**), produced by Smiths, to the trendy KP Discos and the traditional American Corn Chips from Granny Goose of California.

In 1982 Phileas Fogg snacks were launched by Derwent Valley Foods of Consett, England. Their success helped develop a distinctive adult snack market, which includes the spicy Popadoms from KP Foods (a division of the British firm United Biscuits).

1

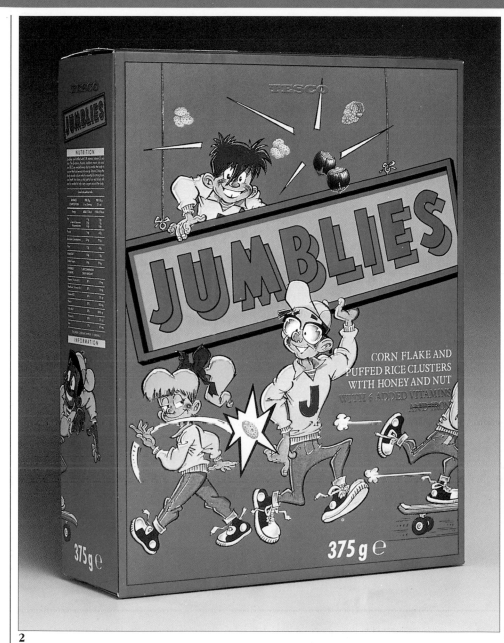

2

3

4

The designs for the "own label" products of the British supermarket chains have found a new vitality in the late 1980s. For instance, in the area of breakfast cereals (many of which are targeted at children), the packs have become more exciting, such as Tesco's Jumblies (2), Marks & Spencer's Breakfast Special (3), and Sainsbury's Puffed Wheat (4).

HOME CARE

The design for Johnson Wax Ltd Lifeguard Disinfectant (*2*) was updated in 1988 using a style reminiscent of advertisements in the late 1920s, when an impression of the image was given rather than any detail.

1

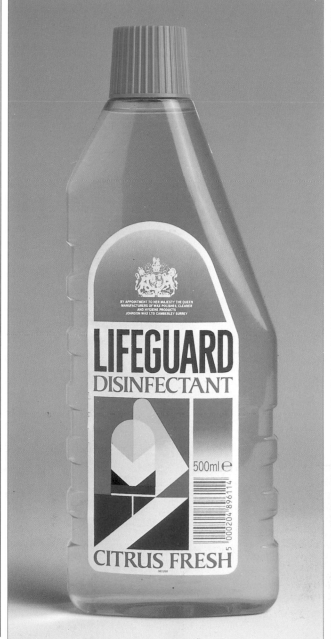

2

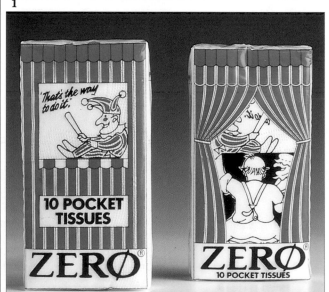

The impression conveyed by the design for Waitrose Washing Powder (*1*) is one of softness. The bubbles and the rippling water flow from the circular emblem of the washing machine.

Zero Pocket Tissues (*3*) are manufactured by Undercover Products International Ltd, Wales. On the back of the pack the puppeteer can be seen manipulating Punch and Judy.

3

4

Another interpretation of a washing powder pack from Coral (4) produced by Lever in Paris. The swirl of the clothes and the colours spin out as coloured balls.

Apart from using bright, modern styles, these packs for Mazda Light Bulbs (6) are colour coded to give instant recognition of the wattage required.

The American product Mr Clean (5) from Procter & Gamble still uses the satisfied smile of the Mr Clean character, created in 1957.

5

6

181

DIY GOODS

Utility metal cans of motor oil have been mostly replaced by utility plastic containers. Now the motorist is offered moulded plastic designer containers to reflect the high performance motor oil provided by British Petroleum (1).

1

In the paint market as well, metal cans as used by Crown (2) are being replaced by plastic, as now used by Dulux (3). Until recently, the colour shade of the paint has been displayed in a dull way; now an imaginative visual image communicates the colour.

2

3

The arrival of the self-service store meant that products needed to be pre-packed and readily accessible. A solution for the awkward product has been to mount it on a card and hang it on a rack. This enables products to be displayed in an orderly fashion largely unaffected by the ravages of customers.

The examples of hanging products shown here have used different methods of achieving the same end: to display the actual object. The electrician's tools from Do It All (**4**) are mounted on a hanging card. The handyman tools from Rabone (**5**) are also card-mounted but are enveloped in clear plastic.

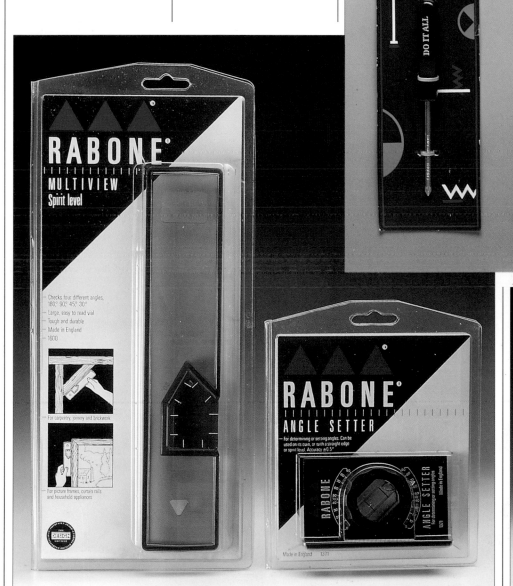

The Durabeam Torch (**6**) from Duracell dispenses with the card backing, using a folded leaflet instead, and relies on the moulded plastic pouch to show off the product. This includes the two batteries, an addition that is usually left to the customer to buy separately.

4

5

6

FRAGRANCES

Production of Joseph Parfum de Jour (**1**) in 1985 has taken the adventurous feel of the spirit flask, making the container an expensive accessory being made from stainless steel. The success of the Body Shop reflects the growing concern for environmental issues. Initially Body Shop products were aimed at women. However, the Mostly Men range (**2**) is the first group of products aimed solely at men. The logo has been designed to run across the whole product range be it after shave in a glass bottle, talc in a cardboard cylinder or shaving cream in a plastic tube.

2

3

This presentation gift set of four Le Jardin (**3**) products by Max Factor displays each item in its own window on the sides of the box.

1

4

5

The box for Summer Blonde (*4*) by Clairol conveys the "sun-drenched look" by means of a stylized girl's face with sun glasses.

The extensive use of moulded plastic bottles has necessitated the development of variety of shapes. In the case of Palmolive Foam Bath (*5*) containing "soft green" foam, the colour pervades the whole product. A self-adhesive label with a clear film has been used, allowing the colour of the plastic to show through.

Even after purchase, the cosmetic needs to look exciting on the dressing table. This has been achieved by the spiralling glass container with a plastic top for Mantana Eau de Toilette spray (*6*).

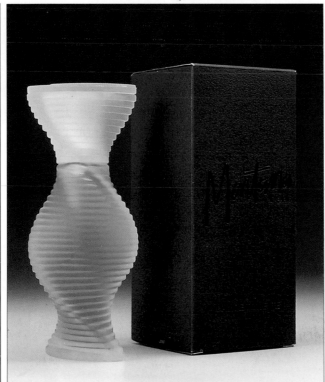

6

7

The use of black and gold aims to give the shaving set, Mancare (*7*), a touch of class. The brand name is obliquely emphasized in the top pattern.

RETROSPECTIVE

There are two ways in which a product may capture the feel of the past. Firstly by taking the style of an earlier period and recreating it on the modern pack; secondly by taking the actual graphics used by that pack in earlier times.

California Classic Pistachios (*1*), made by Panda Foods Corp., have used a sun-burst motif. With the French Petit Nantais biscuits (*3*) the pack has the fussy graphics of an earlier time, although the colouring is from the present day and the biscuit has been reproduced photographically.

1

2
Marks & Spencer have reverted to a pseudo late-Victorian style for their Gravy Granules (2).

3

5

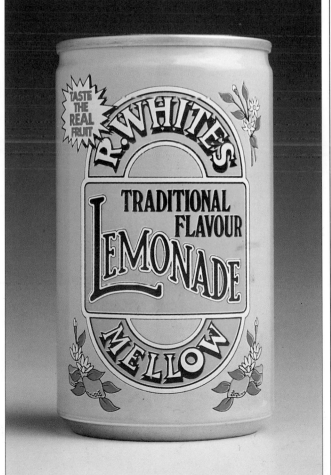

Both Sunlight Soap and R Whites
have been on sale on the British
market for over a hundred years.
In the case of the Sunlight Soap
pack (**5**), the design has reverted to
one based on the original layout,
using similar but modern colours.
With R Whites Lemonade (**6**) the
appeal of traditional flavour has
been complemented with a
traditonal design, although this
was not based directly on any
R Whites graphics of a hundred
years ago.

4

*The classic line for the French
chicory essence Leroux (**4**)
suggests that the product has used
this livery for some time.*

6

INDEX

INDEX

CREDITS

The author would like to thank all the companies and individuals who have helped to supply information for the book and would welcome any further information that readers might supply.

Special thanks are due, in particular, to the following for making material available:
Collection of Phil and Lorna Sarrel: **p21**, 1; **p30**, 1, 3, 4; **p33**, 1, 3; **p34**, 2; **p42**, 3; **p46**, 4; **p47**, 7; **p69**, 5, 6, 7; **p79**, 3, 5; **p104**, 2; **p106**, 1; **p113**, 5.
Collection of Hobie and Nancy Van Deusen: **p54**, 1; **p67**, 3, 4, 5; **p72**, 3; **p88**, 1; **p92**, 3, 4; **p99**, 5; **p108**, 1.

Fisons plc (Gro-bag is a trade name of Fisons plc): **p173**, 3, 4.

Body Shop: **p184**, 2.

Every effort has been made to acknowledge all sources of material. Quarto would like to apologise if any errors or omissions have been made.

The author would like to thank his mother for her continuous support and Polly Powell for her invaluable assistance.